# Sunset Ideas for Cooking
# Vegetables

*By the Editors of Sunset Books
and Sunset Magazine*

LANE PUBLISHING CO. · MENLO PARK, CALIFORNIA

**Edited by Judith A. Gaulke**

**Special Consultant: Annabel Post**
Home Economics Editor, Sunset Magazine

Design : John Flack

Illustrations : Sidney Hoover

Cover Photograph by George Selland, Moss Photography

Editor, Sunset Books : David E. Clark

Tenth Printing June 1979
Copyright © Lane Publishing Co., Menlo Park, CA 94025.
First Edition 1973. World Rights reserved.

Library of Congress No. 72-92516. ISBN 0-376-02903-X.
Lithographed in the United States.

For people who like to grow vegetables in their own home gardens,
a companion to this volume is the Sunset book, Vegetable Gardening.

# CONTENTS

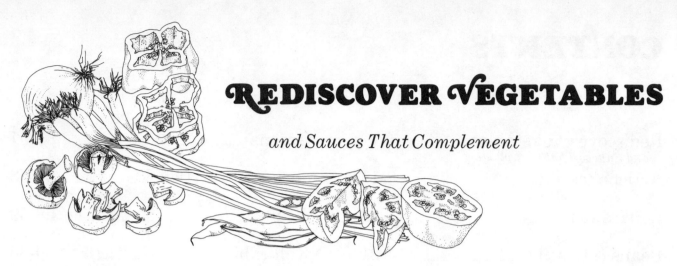

# REDISCOVER VEGETABLES

## *and Sauces That Complement*

Never before have markets offered such variety in vegetables. Even so, a cook can run out of interesting ways to serve them. This book is packed with ideas for cooking vegetables.

It's a cook book for people who love vegetables, for those who would like to get to know them better, and for mothers who wish they could get their families to eat more of them.

However, it's more than a book of recipes. All the information you might need about buying, storing, and preparing vegetables to cook is included for each of the 50 different vegetables. Vegetables are arranged alphabetically so you can turn quickly to whichever one you've brought home from the market. Even if it's a vegetable you have never tried before, you'll find helpful suggestions on how to prepare and serve it.

We emphasize easy ways to add variety in seasonings, cooking techniques that enhance rather than mask the fresh natural flavor of vegetables. Near the beginning of each section, there's a collection of simple serving suggestions. If you've fallen into the habit of cooking vegetables the same way time and time again, it may surprise you to discover how much you can change the flavor simply by cutting and cooking them in a different way. Discover extra sweetness when carrots are thinly sliced and cooked quickly in a frying pan with a few tablespoons each of butter and water. (We call this method *butter-steaming* and give instructions for it under most vegetable categories.)

Does cooking eggplant usually mean frying it in a lot of oil to keep it from sticking to the pan? Try it broiled or charcoal grilled until richly browned; it develops a mellow, caramel-like flavor and requires less oil.

## Help for Planning Menus

Because vegetables can play so many different roles, they become important keys to menu variety. Inspiration might start in the produce section of your market when you see what good buys there are and what looks especially fresh that day. Keep an open mind toward the many forms a well-balanced menu can take, then let your vegetable choices help suggest interesting menus for you.

For example, if you brought home fresh peas (or have some in your garden), look under the heading for *Peas*. You'll find lots of suggestions for serving peas as a hot vegetable as well as for pea soup, chilled fresh pea salads, and suggestions that mix peas with rice or pasta. As you can see, using vegetables in different ways can add many facets to a meal.

## A Word About Company Meals

Most hostesses can enjoy their own parties better when much of the meal has been prepared ahead of time. For vegetables to be tender and still piping hot, they must be cooked during those demanding few minutes just before serving dinner.

One solution to this time pressure is a vegetable casserole that can cook along with the meat in the oven. Another technique is *pre-cooking*. It can be applied to most vegetables. You cut the vegetable carefully into uniform sizes—thin slices, big chunks, or the whole vegetable. Cook them in boiling water or steam, but watch carefully to stop cooking just as soon as they begin to feel tender when pierced with a fork or knife. Then quickly drain (or remove from steamer), plunge into ice cold water to stop cooking completely, and drain well again. Vegetables prepared this way—as far ahead as a day or two—taste remarkably fresh when reheated quickly.

You can choose one of several ways to bring pre-cooked vegetables back to serving temperature. When butter and other seasonings are called for in the recipe, melt the butter and add the seasonings, then add the pre-cooked vegetable. In about the time it takes to gently stir and mix them, the vegetable will be piping hot again and ready to serve. If you let the butter turn a nut-brown color and add some lemon juice and parsley along with the vegetable, then you'll have vegetables in a sauce the French call Beurre Noir. Another variation would be to start by using butter to sauté some onion, garlic, or fresh mushrooms, then stir in the pre-cooked vegetable to create other delicious dishes.

Another way to reheat vegetables is to plunge them into a quantity of simmering water and let them stand just long enough to heat through again; quickly drain and serve.

Sometimes the best answers for vegetables served at guest meals are dishes that taste best at room temperature or chilled. A surprising number of vegetables fall into this category; for example the Mediterranean vegetable stew called Ratatouille (recipe page 48), Marinated Green Beans (recipe page 16), artichokes with Béarnaise (recipe page 9), and vegetables seasoned with olive oil in the Italian or Greek manner. And don't overlook such menu innovations as serving vegetables raw as an appetizer or relish or dressing them with oil and vinegar and letting them double as salad.

## If You are on a Diet

Whether you are watching calories or cholesterol, it is possible to make vegetables taste good without butter or rich sauces. Start with the freshest vegetables available, then cook them with care to preserve all their natural flavor and nutrients. This requires careful timing—an extra minute or two of overcooking can often have a detrimental effect on the taste.

One method of cooking vegetables can be particularly helpful to dieters. If you are familiar with Chinese cooking, you may recognize this technique, for it is similar to the Oriental stir-frying of meat and vegetables.

We have adapted and modernized this Chinese art, making it a quick, easy way to prepare many vegetables. Since the seasonings used are not necessarily Oriental, they fit into almost any meal pattern. When you use a big frying pan or wok (a round-bottomed Oriental pan) over high heat, the vegetables cook quickly in a scant amount of salad oil until just tender, retaining their natural color, flavor, and some crispness.

You'll find directions for *stir-frying* under most vegetables covered in this book. Once you learn to cook vegetables by this method, you'll discover that it is easy to improvise, adding other flavors and creating your own combinations. Since all cutting, measuring, and sometimes pre-cooking is done ahead, it is easy to cook this way for company meals, too. For those times when a sauce is really needed to flavor a vegetable, the dieter can turn to ones with mayonnaise-like bases, such as the Green Herb Sauce, oil and vinegar based Shallot Dressing or sauces based on yogurt or cottage cheese (recipes page 7).

## Switch to Raw Vegetables

Fresh, raw vegetables are appealing to people of all ages, and they make delicious party appetizers that even dieters can appreciate.

For buffets and walk-around parties, a presentation of raw vegetables can serve as both vegetable and salad. One attractive way to present them is to fill a basket with colorful whole vegetables—they can have almost invisible cuts made so they'll break apart easily. Nearby have a bowl of sauce or dressing such as Hollandaise, Home-Made Mayonnaise, Bagna Cauda, or a low-calorie sauce such as Savory Cheese Sauce (recipes pages 6–7).

Here are some vegetables that are delicious when eaten raw, along with suggestions for preparing and presenting them.

*Artichokes*—break off small outer bracts; cut thorny tips from remaining bracts with scissors. Trim stem ends. Keep in acid water (page 8) until ready to serve. To eat, bite off tender base of each bract.

*Cabbage*—cut in half. Cut vertical gashes in each. Break off chunks to eat.

*Carrots*—leave on 1 inch stem. Peel, gash carrot almost through in short sections; break to eat.

*Cauliflower*—cut out core, keep head whole. Break off flowerettes to eat.

*Cherry tomatoes*—dip with stems.

*Green beans*—snap off ends and remove string. Leave whole to eat.

*Green peppers*—cut vertically down to stem in 8 to 12 sections around seed center. Break to eat.

*Mushrooms*—trim stem ends. Eat small mushrooms whole. Cut large ones through cap only into 4 to 6 sections; break to eat.

*Radishes*—cut off root ends and all but 1 or 2 leaves to hold for dipping.

*Turnips*—peel and cut not quite through in thick slices. Break apart to eat.

*Zucchini and yellow crookneck squash*—trim ends; cut almost through in short sections. Break to eat.

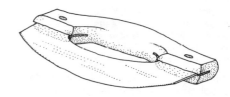

## All-Purpose Vegetable Sauces

No book on vegetables would be complete without some of the classic sauces like Hollandaise, Béarnaise, Mornay, and the not quite so well known Italian Pesto Sauce. They complement many vegetables and are often referred to throughout this book.

## Hollandaise Sauce

2 egg yolks
3 tablespoons warm water
1 tablespoon lemon juice
¾ cup (⅜ lb.) melted butter

In the top of a double boiler, put the egg yolks, warm water, and lemon juice and beat together with a wire whip or rotary beater. Place pan over gently simmering water (water should not boil or touch bottom of pan) and beat constantly as you gradually add the melted butter in a slow, steady stream. Continue to beat and cook after all butter is added until sauce thickens slightly. Remove sauce from hot water immediately and serve; or let stand at room temperature for as long as several hours.

*To bring Hollandaise back to serving temperature*, set pan in water that is just hot to the touch and stir occasionally for 5 to 10 minutes until lukewarm. Makes 1 1/3 cups (enough for 4 to 6 servings).

**Blender Hollandaise Sauce.** Combine 3 egg yolks (at room temperature) and 1 1/2 tablespoons lemon juice in the jar of the blender. Melt 3/4 cup (3/8 lb.) butter and heat until it bubbles (don't brown). Add 1 tablespoon hot water to egg mixture, turn blender on high speed, and immediately pour in the hot butter in a slow, steady stream. Whirl until well blended. Turn off blender and serve.

For added flavor, try adding 1/2 teaspoon salt, a dash cayenne, and 1 teaspoon prepared Dijon-style mustard after butter is incorporated. Whirl 30 seconds more and serve.

**Tart Hollandaise Sauce.** For this variation, combine in the blender 3 egg yolks, 3 tablespoons lemon juice, and 1/4 teaspoon dry mustard. Whirl at high speed for about 30 seconds and with blender motor running, gradually pour in 1/2 cup hot melted butter then add 1/4 cup cold butter (cut in small pieces) for a total of 3/4 cup (3/8 lb.) butter. Whirl again and serve.

## Mornay Sauce

1 tablespoon *each* butter or margarine and flour
  Dash paprika
¼ teaspoon salt
1 cup half-and-half (light cream)
¼ cup milk
¼ cup shredded Gruyère or Samsoe cheese
2 tablespoons shredded Parmesan cheese
½ teaspoon prepared Dijon-style mustard
  Dash white pepper
  Drop liquid hot pepper seasoning

In a saucepan melt the butter. Stir in the flour, paprika, and salt; cook until bubbly. Then gradually stir in cream and milk, cooking over low heat, stirring constantly, until slightly thickened. Stir in the cheeses, mustard, pepper, and liquid hot pepper seasoning.

Continue cooking over low heat until thickened and smooth, stirring frequently, about 5 minutes. Serve immediately or let cool; store, covered, in refrigerator.

To reheat, place in top part of double boiler over gently simmering water; stir occasionally until heated through (about 5 minutes). Makes 1 1/2 cups sauce.

## Béchamel Sauce

¼ cup (⅛ lb.) butter or margarine
¼ cup all-purpose flour
1 cup regular strength chicken broth
1 cup half-and-half (light cream)
  Salt and pepper to taste
  Dash ground nutmeg (optional)

In a saucepan or in a double boiler over simmering water, melt butter and stir in flour until well blended and bubbly (do not let butter brown). Remove from heat. Gradually stir in chicken broth and cream. Cook, stirring constantly, until mixture is thickened and smooth. Add salt, pepper, and nutmeg to taste.

## Béarnaise Sauce

1 egg
1 teaspoon *each* tarragon vinegar and Dijon-style mustard
1 tablespoon lemon juice
½ cup warm, melted butter

Place the egg, vinegar, mustard, and lemon juice in the container of the blender. Whirl a few seconds, then with motor running, slowly pour in butter, blending until thickened. Keep warm in top of double boiler over hot water until ready to use; or make ahead and reheat slowly in double boiler over barely simmering water, stirring occasionally, until smooth. Makes about 3/4 cup.

## Basic Pesto Sauce

Pesto is made from fresh basil leaves and can be purchased from the freezer sections of many markets. It is also easily made at home. Use it directly on vegetables to season or blend it with butter, into mayonnaise, or into a dressing.

2 cups packed fresh basil leaves, washed and drained well
1 cup freshly grated Parmesan cheese
½ cup olive oil

Put basil in a blender jar; add cheese and oil and cover. Turning motor on and off, whirl at high speed until a very coarse purée; push pesto down from sides frequently with a spatula. Use at once, cover, and refrigerate up to a week; freeze in small portions. Surface will darken when exposed to air, so stir before using. Makes about 1 1/2 cups.

**Pesto Butter.** Blend 3 tablespoons Basic Pesto Sauce with 1/2 cup soft butter. Add to hot cooked artichokes, green beans, broccoli, carrots, peas, or a variety of other vegetables. Makes 2/3 cup butter.

**Pesto Dressing.** Blend 6 tablespoons Basic Pesto Sauce with 1/3 cup wine vinegar, 2/3 cup olive oil, and 1 minced garlic clove. Mix. Use as a marinade for uncooked vegetables such as sliced mushrooms, tomatoes, cucumber, or zucchini. Makes 1 1/3 cups.

**Pesto Mayonnaise.** Combine 1 egg, 2 tablespoons lemon juice, and 1 clove garlic in blender jar. Cover and whirl at high speed until blended. Turn motor off, add 1/2 cup Basic Pesto Sauce, then turn on high speed and gradually pour in 1/2 cup melted butter and 3/4 cup salad oil. For a sauce that has pouring or dipping consistency, serve at once or at room temperature; for a thicker sauce, cover and chill. Use with any vegetables mentioned above or serve as a dip for hot or cold whole artichokes. Makes 2 cups.

## Homemade Mayonnaise

1 egg
½ teaspoon *each* sugar and paprika
2 teaspoons prepared Dijon-style mustard
3 tablespoons tarragon-flavored white wine vinegar
1 cup salad oil

Place in a blender the egg, sugar, paprika, mustard, and vinegar. Blend a few seconds and, with motor running, gradually pour in salad oil, blending until smooth. Chill. Makes 1 1/2 cups.

## Bagna Cauda Sauce

Use as a dipping sauce for raw vegetables suggested under *A Word About Raw Vegetables*, page 5.

½ cup (¼ lb.) butter
¼ cup olive oil
4 small cloves garlic, mashed
1 can (2 oz.) flat anchovy fillets, drained well

Choose a heatproof container that will be only about half filled by the quantity of sauce you make. Combine the butter, olive oil, and garlic. Finely chop anchovy fillets and add to sauce. Stir over moderate heat until mixture bubbles.

To serve, set over candle or low alcohol flame. Mixture must not get hot enough to brown and burn. Makes 8 to 10 servings (double recipe for 16 to 20 servings; triple for 24 to 30).

## Shallot Dressing

For those times when you would like to serve vegetables like a salad, this is delicious with both raw and lightly cooked vegetables. Try it on tomato slices, raw mushroom slices, or thinly sliced raw zucchini. It is also good with cooked and thinly sliced beets or carrots and French cut green beans cooked until just limp. Vegetables are best served at tepid temperatures.

⅓ cup thinly sliced shallots, or green onions, including tops
4 teaspoons prepared Dijon-style mustard
¼ cup white wine vinegar
⅛ cup salad oil or olive oil
¼ teaspoon salt
⅛ teaspoon pepper
Raw or cooked vegetables

Blend together the shallots or green onions, mustard, vinegar, oil, salt and pepper; stir or shake vigorously just before adding to vegetables. You can cover and keep dressing up to 2 days. Makes about 1 cup dressing.

## Savory Cheese Sauce

Serve this as a topping for baked or broiled potatoes, over carrots, green beans, and Brussels sprouts. As an appetizer it makes a great dip for artichokes or other raw vegetables.

1 cup cottage cheese
1 slice of medium-sized onion (about ¼ inch thick)
1 tablespoon lemon juice
½ cup unflavored yogurt
1 teaspoon sugar
  About ½ teaspoon salt
  Dash pepper
  About ½ teaspoon dill weed or *fines herbs*
1 to 2 tablespoons finely chopped parsley

Put into the blender container the cottage cheese, onion, lemon juice, and 3 tablespoons of the yogurt. Whirl until smooth and mounding like sour cream, stopping the motor and pushing cheese down into blades as needed. Turn into a bowl and stir in remaining 5 tablespoons yogurt, sugar, salt, pepper, dill weed or herbs, and parsley. Cover and store in refrigerator. Makes about 1 1/2 cups.

## Green Herb Sauce

This sauce goes particularly well with broccoli, artichokes, carrots, and baked potatoes.

½ cup watercress leaves and small stems (or spinach leaves), pressed in cup
½ cup parsley sprigs, pressed in cup
1 large shallot, peeled and cut (or 1 large green onion with top), sliced
½ teaspoon *each* tarragon and thyme leaves (or 1 teaspoon *fines herbes*)
½ teaspoon salt
¾ teaspoon dry mustard
2 tablespoons white wine vinegar (or tarragon vinegar)
1 egg
1 cup salad oil
  Raw or cooked and cooled vegetables

Put into the blender container the watercress or spinach, parsley, shallot or onion, herbs, salt, mustard, vinegar, and egg. Whirl until liquified. With blender motor running, remove center of top and begin pouring oil in a thin stream; add oil very slowly in the beginning, then a little faster as it begins to thicken. Store in refrigerator, tightly covered. Makes about 1 2/3 cups.

# ARTICHOKES

## Selection and Storage

**Selection:** Choose artichokes with tight, compact heads (they are actually flower buds). If leaves have opened out, they are over-mature and may be woody; if leaves are curling in on sides and brown, they are too old. Surface brown spots during winter months, however, indicate the bud was touched by frost; this does not harm flavor. Size is not indication of quality. The largest buds grow at tip of plant, and those that form farther down stem are decreasingly smaller in size. The smallest (2 1/2 inches or less in diameter) are often called hearts.

**Buying:** Allow 1 medium or large size whole artichoke for each serving. Large ones have a 4-inch diameter or more, medium-sized are 2 3/4 to 3 3/4 inches in diameter, and small are 2 1/2 inches in diameter or less.

**Storage:** Keep unwashed and dry, tightly wrapped or in plastic bags in refrigerator. Good artichokes stay fresh for up to 2 weeks.

## Cutting and Cooking

Cut artichokes darken when exposed to air, so prepare them just before cooking. If you are preparing several artichokes at a time, drop into an *acid-water bath* (3 tablespoons vinegar mixed with each quart water) to cover artichokes as each is cut. Avoid using carbon steel knives or cast iron or rolled steel pans because they discolor artichokes and may give them a metallic taste.

**To trim whole artichokes,** slice off the thorny tip. Then with scissors, cut thorns from tips of remaining leaves. Remove small leaves around base and peel stem (you can cut off stem, making flat base on bottom before or after cooking).

**When you want to use artichokes as containers** for sauce or cases for stuffing, first slice off top third of each, then reach in with a spoon and scrape out fuzzy center and small center leaves. (Or you can cut out centers after cooking.)

**For completely edible artichokes,** slice off top third of each. With a spoon remove center choke and small inside leaves. Break off all coarse outer leaves down to pale green inner leaves and peel

the stem. Use whole or cut in halves or quarters. Prepare small artichokes (called hearts) this way, leaving centers intact. Rinse artichokes in cool water just before cooking.

**For boiled artichokes,** 4 large whole artichokes (or 6 to 8 medium or 36 small artichokes), you will need a 5-quart or larger kettle. Bring to a boil in kettle 4 quarts water with 2 teaspoons salt. If desired, add 4 to 6 tablespoons vinegar, 3 to 4 tablespoons olive oil, 2 bay leaves, and 10 to 12 whole black peppers.

Add trimmed artichokes to boiling liquid and place a plate on top of water to keep artichokes submerged. Cover pan, return to a boil, and cook until stem end pierces readily with a fork.

Allow 15 to 20 minutes for small artichokes (hearts), about 25 to 35 minutes for medium-sized artichokes, and 40 to 50 minutes for large artichokes.

Lift from cooking liquid, drain, and serve hot with melted butter, mayonnaise, or cold with any of the sauce recipes that follow. Or use as directed in recipes calling for cooked artichokes.

If you want to cook artichokes ahead and reheat them, return the drained artichokes to simmering liquid for 5 to 10 minutes or until heated through.

**To pressure cook whole artichokes,** follow directions under *cutting and cooking* (at left) for trimming whole artichokes.

Place 1 to 4 medium to large artichokes in a 4 to 8-quart pressure cooker (same quantity water should be used regardless of number of artichokes used). Following manufacturer's directions for operation, add 1 cup water for a 4-quart cooker, 1¼ cups for a 6 to 8-quart cooker. Cook 15 minutes under 15 pounds pressure after control jiggles.

Reduce pressure by placing under cold running water 15 seconds. Remove lid. Take out artichokes and drain well. Serve hot or cold with any of the sauces (recipes follow).

**To eat a cooked whole artichoke,** pull off leaves one by one and dip the base of each into a sauce. Eat only the tender light green base part of each leaf by drawing it between your teeth. Discard the remaining fibrous part of the leaf (tip). Continue pulling off leaves until you come to the fuzzy center part called the "choke." Scoop it out with a spoon and discard. Cut the succulent base that remains (called the artichoke bottom) into bite-sized pieces and eat with a fork.

## Artichoke Sauces

Follow directions under *cutting and cooking* (page 8) for trimming whole artichokes; also scoop out centers if you want to serve artichokes with sauce inside. Cook in boiling water or pressure cook (directions precede); drain well.

### Cream Cheese Béarnaise

In a small saucepan, boil 2 tablespoons tarragon vinegar with 2 tablespoons minced green onions and 1/4 teaspoon tarragon leaves until vinegar is evaporated. Combine onion mixture with 2 small packages (3 oz. *each*) chive-flavored cream cheese, 2/3 cup freshly grated parmesan cheese, and 2 tablespoons lime juice; mash with a fork until smoothly blended. Thin to a consistency that can be dipped but is not runny by adding 1 teaspoon milk or whipping cream at a time. Serve with hot cooked artichokes or chill the sauce, then whip lightly and serve with chilled artichokes. Makes about 1 cup.

### Egg Sauce

In a small bowl, mash 1 hard-cooked egg with a fork. Thoroughly blend in 3/4 cup sour cream, 1/4 cup mayonnaise, and 1/2 teaspoon *each* dill weed and salt. Serve with chilled artichokes. Makes enough for 6 servings.

### Curry Mayonnaise

Blend about 1 teaspoon curry powder with 1 cup mayonnaise. Serve with chilled artichokes. Makes enough for 6 servings.

### Horseradish Sour Cream Sauce

In a small bowl, thoroughly blend 1/4 teaspoon prepared horseradish (or more to taste) and 1/2 teaspoon salt with 1/2 pint (1 cup) sour cream. Serve with chilled artichokes. Makes enough for 6 servings.

### Artichokes with Poached Eggs and Tart Hollandaise

Trim 6 large or medium-sized artichokes as directed under *cutting and cooking*, page 8, cutting off top and hollowing out center to use as containers. Cook in boiling water and drain well. Poach 6 eggs until softly set. Lift eggs from water with a slotted spoon, drain briefly, then slip one into each artichoke. Pour 2 or 3 tablespoons warm Tart Hollandaise Sauce (recipe page 6) into each and serve. To eat, pull off leaves, dipping them in sauce; then with a fork eat eggs. Makes 6 servings.

## Serving Suggestions

Artichokes need not always be presented whole and with a sauce. Try cutting artichokes down until they are completely edible and cooking them quickly in a frying pan (see recipes below).

### Butter-Steamed Artichokes

Trim 12 to 18 small artichokes (called hearts) as directed under *cutting and cooking* on page 8 to make completely edible—do not remove center chokes. Cut each in quarters (you should have 4 to 5 cups), or use thawed frozen artichokes. In a 10-inch or larger frying pan that has a cover, melt 2 tablespoons butter or margarine over high heat. Add the artichokes and 6 tablespoons water; stir. Cover and cook on high heat, stirring occasionally, until tender, about 8 minutes (about 6 minutes for frozen). Season to taste with salt and serve immediately. Makes 4 to 6 servings.

### Artichokes with Tarragon

Prepare Butter-Steamed Artichokes as above, adding 1/2 teaspoon tarragon leaves along with the artichokes. Cook as directed. Season with salt and 2 tablespoons lemon juice.

### Cream Glazed Artichokes with Basil

Prepare Butter-Steamed Artichokes as above, adding 1/2 teaspoon basil leaves along with the artichokes. After cooking for 8 minutes, remove cover, add salt to taste, and 4 to 5 tablespoons whipping cream. Cook, stirring, until liquid is almost gone.

## Easy Artichokes

1 package (8 to 9 oz.) frozen artichoke hearts
1 tablespoon butter or margarine
⅛ teaspoon basil leaves
   Dash salt and pepper

Cook frozen artichoke hearts according to package directions; drain. Season with butter, basil, salt, and pepper; cover and keep warm. Makes 2 to 3 servings.

## Artichoke Hearts Italiano

1 package (8 to 9 oz.) frozen artichoke hearts
   About 3 ounces jack cheese, sliced
⅛ teaspoon *each* salt, pepper, and oregano leaves
1 can (about 7¾ oz.) marinara sauce

Cook artichoke hearts according to package directions. When tender, drain well and place in a shallow 1-quart baking dish. Arrange cheese slices on top, sprinkle with salt, pepper, and oregano, and pour the sauce over the top. Bake, uncovered, at 350° until sauce bubbles (15 to 20 minutes). Serve artichokes immediately. Makes 3 to 4 servings.

## Marinated Artichoke Hearts

4 pounds small artichokes (*each* about 2 inches in diameter)
1 cup *each* olive oil and white vinegar
1 *each* whole carrot, small onion, and celery stalk
2 cloves garlic
1 stick whole cinnamon
5 bay leaves
½ teaspoon *each* whole black peppers and salt

Trim artichokes so they are completely edible, following directions under *cutting and cooking*, page 8 (do not remove centers); be sure to use *acid-water bath* when preparing this quantity. When all artichokes are trimmed, drain from acid-water bath and put artichokes in a large pan. Add oil, vinegar, carrot, onion, celery, garlic, cinnamon, bay leaves, whole peppers, and salt. Cover and bring to a boil, reduce heat, and simmer until tender, about 20 minutes.

Let stand in cooking liquid, covered, overnight. Remove onion, bay leaves, and celery. Cut artichokes in half or leave very small ones whole. Return to marinade and bring to a boil. With slotted spoon, lift artichokes from liquid; arrange in 2 wide-mouth pint jars or in a bowl. To each pint, add 1 or 2 slices of the carrot, half the cinnamon stick, and 1 clove garlic. Bring the remaining marinade to a boil. Cover the artichokes with the liquid. Store covered in the refrigerator. Makes 2 pints.

## Artichokes with Mushrooms

¾ pound mushrooms, halved
4 tablespoons (⅛ lb.) butter or margarine
2 packages (8 to 9 oz. *each*) frozen artichoke hearts
⅓ cup whipping cream
½ teaspoon tarragon leaves
Salt and pepper to taste

In a saucepan, sauté mushrooms in the butter for 5 minutes over medium heat. Add the frozen artichoke hearts. Cover and simmer 7 or 8 minutes longer. Stir in the cream, tarragon, salt, and pepper. Makes 4 to 6 servings.

## Crisp Artichoke Appetizers

5 medium-sized artichokes
1 egg
¼ cup water
½ teaspoon salt
¼ teaspoon pepper
About 2 cups salad oil
Bread crumbs

Wash artichokes; drain. In a bowl, beat egg, water, salt, and pepper. Cut off top half of each artichoke; trim stem to 1 inch. Snap off tough outer leaves down to pale green leaves; trim base. Slice artichoke in half lengthwise; remove fuzzy choke. Then holding each half

with steady hand, carefully slice 1/4-inch cross sections from each; cut end pieces in half lengthwise.

In a 2-quart sauce pan or deep fat fryer heat the oil to 375°. Dip each artichoke slice into egg mixture, then coat in bread crumbs and fry in oil until golden brown, turning once (takes about 1 minute). Remove with slotted spoon, drain well, and serve hot. Makes 2 1/2 to 3 dozen appetizers.

## Artichoke Halves with Crab

5 to 6 medium-sized artichokes
1 tablespoon butter or margarine
1½ tablespoons all-purpose flour
⅔ cup regular strength chicken broth
½ cup half-and-half (light cream) or milk
½ teaspoon basil leaves
1 pound crab meat
¾ cup shredded Swiss cheese

Trim artichokes following directions under *cutting and cooking*, page 8, to make them completely edible; peel stems but leave them in place. Cut artichokes in halves lengthwise and cook in boiling water as directed. Drain well, and with a small spoon scoop out choke and tiny inner leaves.

While artichokes are cooking, melt butter in a saucepan and blend in flour. Gradually add chicken broth, half-and-half, and basil; cook, stirring, until boiling and slightly thickened. Remove from heat and mix in crab meat.

Pour crab mixture into a 1 to 1 1/2-quart shallow casserole. On top, arrange the cooked artichoke halves side by side, cupped side up. Sprinkle with Swiss cheese, covering artichokes well to prevent drying. Bake, uncovered, in a 400° oven for about 12 minutes or until bubbling. (To make dish ahead, assemble, cover, and chill; uncover and bake for about 20 minutes or until bubbling.) Makes 4 to 5 servings.

## Spinach with Artichokes au Gratin

2 jars (6 oz. *each*) marinated artichokes
3 packages (10 oz. *each*) frozen chopped spinach, thawed
3 small packages (3 oz. *each*) cream cheese
4 tablespoons soft butter or margarine
6 tablespoons milk
Pepper
⅓ cup shredded Parmesan cheese

Drain marinade from artichokes and save for other uses. Reserve a few artichokes to use later as garnish; distribute the remainder over the bottom of a shallow 1 1/2-quart casserole.

Squeeze as much moisture as possible from spinach and arrange the chopped leaves evenly over artichokes. With a mixer, beat cream cheese and butter until smooth and fluffy, then gradually blend in the milk. Spread this mixture over the spinach, sprinkle lightly with pepper, then dust with Parmesan cheese. Cover and refrigerate for as long as 24 hours.

Bake, uncovered, in a 375° oven for 40 minutes or until topping is light brown and vegetables are hot. Garnish with reserved artichokes. Makes 10 to 12 ample servings.

## Artichoke Frittata Sandwiches

1 package (8 to 9 oz.) frozen artichoke hearts
  Boiling salted water
3 tablespoons butter or margarine
2 green onions, including part of tops chopped
8 eggs
2 teaspoons salt
  Pepper
8 teaspoons water
½ cup shredded Parmesan cheese
4 slices rye bread
4 leaves lettuce

Cook artichoke hearts in boiling salted water for 5 to 7 minutes or until tender; drain. Add 1 tablespoon butter and the chopped green onions. Heat, stirring, 1 minute.

For each sandwich, beat 2 eggs just until blended with 1/2 teaspoon salt, dash of pepper, and 2 teaspoons water. Heat a 7 or 8-inch frying pan and add 2 teaspoons butter. When it stops sizzling, pour in the egg mixture. Cook over medium-high heat until the mixture sets on the bottom; add 1/4 of the artichokes and onions and sprinkle with 2 tablespoons of the cheese. Place under a preheated broiler, about 4 inches from the heat, until cheese melts and the top is slightly browned. Slip onto a platter and place in a warm oven while you make 3 more frittatas. Toast and lightly butter the rye bread,

cover with lettuce leaves, and arrange a frittata on top of each. Makes 4 open-faced sandwiches.

## Artichoke Nibbles

2 jars (6 oz. *each*) marinated artichoke hearts
1 small onion, finely chopped
1 clove garlic, minced or mashed
4 eggs
¼ cup fine dry bread crumbs
¼ teaspoon salt
⅛ teaspoon *each* pepper, oregano leaves, and liquid hot
  pepper seasoning
½ pound sharp Cheddar cheese, shredded (about 2 cups)
2 tablespoons minced parsley

Drain marinade from 1 jar of the artichokes into a frying pan. Drain the other jar (save marinade for other uses). Chop all the artichokes; set aside. Add onion and garlic to frying pan and sauté until onion is limp, about 5 minutes.

In a bowl, beat the eggs with a fork. Add the crumbs, salt, pepper, oregano, and hot pepper seasoning. Stir in the cheese, parsley, artichokes, and onion mixture. Turn into a greased 7 by 11-inch baking pan. Bake in a 325° oven for about 30 minutes or until set when lightly touched. Let cool in pan, then cut into 1-inch squares. Serve cold or reheat in the pan in a 325° oven for 10 to 12 minutes. Makes about 6 dozen appetizers.

# ASPARAGUS

## Selection and Storage

**Selection:** Choose asparagus with tightly closed, compact tips and firm, brittle stalks that are green almost entire length (except for white asparagus, produced by growing stalks underground). Diameter of stalks is not an indication of tenderness—fat, chunky spears, from beds that are over 5 years old, are generally considered most choice.

**Buying:** Allow about 1/2 pound for each serving.

**Storage:** Asparagus should not be rinsed. Wrap stem ends of stalks in wet paper towels; seal inside a plastic bag and keep in refrigerator. Use as soon as possible.

## Cutting and Cooking

To snap off tough end, grasp stalk with both hands and bend with gentle pressure. Rinse thoroughly. If the scales are gritty, scrape them off.

**Cooked whole spears:** In a large shallow pan, such as a frying pan, lay spears parallel (no more

than 2 or 3 layers deep) in enough boiling salted water to cover. Cook over high heat until water resumes boiling, then simmer just until asparagus is easy to pierce with tip of a sharp knife (about 8 minutes); drain at once. (Or stand stalks upright in a tall pan, coffee pot, or asparagus steamer in boiling salted water and cook until just tender.)

**Boiled asparagus slices:** Cut each spear into thin, slanting slices or into slanting slices that are each about 1 1/2 inches long. Drop into boiling salted water to cover; cook rapidly until just tender when pierced with knife tip, 2 to 5 minutes. Drain immediately.

## Seasoning and Serving

Fresh asparagus cooked carefully to the point when it is bright green and just tender when pierced with a fork is delicious seasoned simply with melted butter. Hollandaise Sauce (recipe page 6) is another classic seasoning for asparagus. Or for variety, try any of the ways given below to add flavor; each recipe is sufficient for about 2 pounds fresh asparagus or 2 packages (9 or 10 oz. *each*) frozen asparagus. Trim and cook fresh asparagus, whole or sliced, following directions under *cutting and cooking*, on page 11; cook the frozen asparagus as directed on package.

## Asparagus with Onion Butter

Combine 1/4 cup butter with 1 teaspoon instant minced onion in a small pan; cook until browned. Add a dash of Worcestershire and pour over hot cooked asparagus.

## Asparagus with Cashew Butter

Melt 1/4 cup butter in a small pan. Add 2 teaspoons lemon juice, 1/4 teaspoon marjoram leaves, and 1/4 cup salted cashews, coarsely chopped. Simmer over low heat for 2 minutes; pour over hot cooked asparagus.

## Asparagus Polonaise

Melt 1/4 cup butter in a small pan, add 2 tablespoons soft bread crumbs, and sauté until lightly browned. Remove from heat and mix in 1 hard-cooked egg yolk, pressed through a wire strainer, and 1 tablespoon minced parsley. Spoon over hot, cooked asparagus.

## Asparagus Vinaigrette

In a jar or bowl, combine 1/2 cup olive oil or salad oil with 2 tablespoons wine vinegar, 1/2 teaspoon salt, and generous dash of freshly ground pepper. Shake or beat to blend and pour over hot, cooked asparagus.

## Asparagus with Chive Sauce

In a small saucepan, combine 1 package (3 oz.) chive cream cheese and 1 tablespoon milk. Stir over low heat until warm and blended. Pour over hot, cooked asparagus.

## Asparagus with Sour Cream Topping

In a small pan, blend 1 cup sour cream with 1 teaspoon prepared mustard, 2 teaspoons lemon juice, and 1/4 teaspoon salt; heat and stir until warm. In a frying pan, melt 2 tablespoons butter or margarine, add 1/2 cup soft bread crumbs, and sauté until browned. Spoon the sour cream mixture over hot cooked asparagus, then sprinkle with the buttered crumbs.

## Asparagus Baked in Wine Sauce

Arrange the cooked spears (they may be hot or cold) in a shallow buttered baking dish. In a pan, melt 1/4 cup butter, stir in 1/4 cup white wine, 1/2 teaspoon salt, and 1/4 teaspoon pepper; pour over asparagus. Sprinkle with 1/3 cup freshly grated Parmesan cheese. Bake, uncovered, in a 425° oven for about 15 minutes (about 25 minutes if refrigerated).

## Chilled Asparagus and Browned Butter Mayonnaise

Trim 2 pounds asparagus spears to the same length (save whatever tender ends you trim to use raw and thinly sliced in green salads). In a wide shallow pan, bring to a boil 3 cups beef broth and season with 1 thinly sliced carrot, 1 thinly sliced lemon, 3 sprigs parsley, 1 tablespoon diced onion, and 1 teaspoon salt. Add asparagus, keeping spears parallel, and cook covered just until tender. Chill in stock. Drain spears and place upright in a bowl or flat on a tray; serve with browned butter mayonnaise as appetizer and dip or as vegetable and sauce.

*Browned Butter Mayonnaise.* Heat 1/2 pound (1 cup) butter until richly browned. Remove from heat and let cool slightly. Beat 4 egg yolks until thick. Add warm butter 2 tablespoons at a time, beating constantly with a rotary beater (or whirl yolks in blender, then add butter in a slow, steady stream). Serve immediately or chill and whip to soften before serving. Makes about 1 1/2 cups.

## Baked Asparagus, Italian Style

Arrange the cooked spears (they may be hot or cold) in a shallow buttered baking dish. Sprinkle with 3 tablespoons freshly grated Parmesan cheese. Pour evenly over cheese 3 tablespoons melted butter. Bake, uncovered, in a 450° oven for about 5 minutes or until lightly browned.

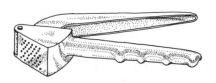

## Butter-Steamed Asparagus

Snap off tough ends from 1 pound asparagus and cut in thin slices. Use a 10-inch or larger frying pan that has a cover or electric frying pan. Melt 2 tablespoons butter or margarine in the pan over high heat. Add asparagus and 2 tablespoons water; stir. Cover and cook at high heat, stirring occasionally, until spears are tender and liquid is gone (2 to 3 minutes). Season to taste with salt and serve immediately. Makes 2 to 3 servings.

## Asparagus with Pine Nuts and Cheese

Follow directions for Butter-Steamed Asparagus above. Add 1/4 cup pine nuts along with the salt and cook, stirring, for about 30 seconds longer. Pour into a serving bowl and sprinkle with 1/2 cup shredded Swiss cheese; stir to blend. Makes 3 to 4 servings.

## Stir-Fried Asparagus

About 2 pounds asparagus
Boiling salted water
1 tablespoon salad oil
1 or 2 cloves garlic, minced or mashed
1 teaspoon finely chopped fresh ginger (optional)
1 tablespoon water
1 teaspoon sugar
About ½ teaspoon salt
Chopped parsley or sliced green onion tops (optional)

Snap off tough ends from asparagus and cut in thin slanting slices. Drop into a large quantity boiling salted water and pre-cook until just tender when pierced with a fork (about 2 minutes). Drain immediately, cool, and set aside. Before starting to cook, prepare all remaining ingredients and have within reach of your range. Heat a 10-inch or larger frying pan or wok over high heat and put in oil. As soon as oil is hot enough to ripple when pan is tipped, put in garlic and ginger, if used; quickly stir and turn with a wide spatula until it starts to brown (about 30 seconds). Put in asparagus, water, sugar, and salt. Keep turning with spatula until heated through (about 1 minute). Turn out on warm serving dish and garnish with parsley or onion, if desired. Makes about 4 servings.

## Asparagus Salad Supreme

2 cans (10½ oz. each) asparagus tips
2 envelopes unflavored gelatin
1 can (10½ oz.) condensed cream of asparagus soup
½ cup dry white wine
1 tablespoon lemon juice
1 cup each sour cream and finely diced radishes
3 tablespoons chopped green onion tops

Drain liquid from asparagus and measure 1/2 cup. Soften gelatin in the canned asparagus liquid. Discard any remaining liquid; set asparagus aside. Mix the soup, wine, and gelatin mixture; heat to just under boiling, stirring until smoothly blended and gelatin is dissolved. Remove from heat and add lemon juice.

Chill until consistency of honey; then fold in sour cream, radishes, and green onions. Decoratively arrange about half of the asparagus tips in a 1 1/2-quart mold; chop remaining asparagus and fold into gelatin mixture. Pour gelatin into mold and chill until firm. Unmold on serving plate. Makes 8 to 10 servings.

## Asparagus and Rice Casserole

3 cups water
1 teaspoon salt
1 pound asparagus, tough ends removed
2 tablespoons butter or margarine
1 each small onion and garlic clove, minced
1 cup long grain rice
1 small jar (2 oz.) sliced pimientos
½ cup grated Parmesan cheese

Bring water to a boil, add salt and asparagus, and cook until asparagus is just tender; drain, reserving the cooking liquid. In a frying pan, heat the butter; add onion and sauté until golden. Add the garlic and rice and stir until rice is milky. Stir in the asparagus cooking liquid. Pour into a 2-quart casserole. Cover and bake in a 400° oven for 15 minutes.

Cut top 2 inches from asparagus and reserve. Chop asparagus ends, and stir into rice with pimiento. Add more hot water, if needed, to keep rice moist. Return to oven for 20 minutes more. Shortly before serving, stir in half the cheese; taste and add salt, if needed. Arrange asparagus tips on top; sprinkle with remaining cheese. Put back into a 400° oven, uncovered, for about 10 minutes. Makes 6 to 8 servings.

## Asparagus Eggs Amandine

2 pounds asparagus
6 to 8 hard-cooked eggs, sliced
4 tablespoons (⅛ lb.) butter or margarine
¼ cup minced onion
2 teaspoons curry powder
2 tablespoons all-purpose flour
2 cups (1 pt.) half-and-half (light cream)
¾ teaspoon salt
¼ teaspoon Worcestershire
⅛ teaspoon pepper
2 tablespoons minced parsley
½ cup sliced almonds

Cook whole asparagus spears as directed under *cutting and cooking*, page 11. Drain asparagus and arrange in a heatproof serving platter. Arrange sliced eggs over asparagus, cover dish with foil, and set in 225° oven.

In a saucepan, heat 3 tablespoons of the butter; sauté onion until tender. Stir in the curry and flour and cook until bubbly. Gradually stir in cream and cook, stirring, until thickened. Stir in salt, Worcestershire, pepper, and parsley; keep warm. Lightly toast almonds in remaining 1 tablespoon butter. Just before serving, pour sauce over asparagus and eggs and sprinkle with almonds. Makes about 4 servings.

## Fresh Asparagus Bisque

2 pounds asparagus
2 tablespoons butter or margarine
4 green onions, including part of tops, sliced
1 small potato, peeled and diced
1 large can (47 oz.) regular strength chicken broth
½ teaspoon salt
⅛ teaspoon pepper
½ teaspoon each Worcestershire and dill weed
2 egg yolks, beaten
½ cup whipping cream or half-and-half (light cream)

Snap off tough ends of the asparagus and discard. Rinse asparagus spears well and cut into 1-inch pieces; set aside.

In a Dutch oven or other large pan, melt the butter over medium heat. Add the green onion and cook until limp (about 3 minutes). Add the potato, chicken broth, salt, pepper, Worcestershire, dill weed, and asparagus. Cover and simmer until vegetables are very tender (about 30 minutes). Remove from heat and pour the soup into a food mill, pressing vegetables through (or whirl part of soup at a time in a blender).

Return soup to pan. Blend egg yolks with cream; stir into soup and heat, stirring until hot. Serves 6.

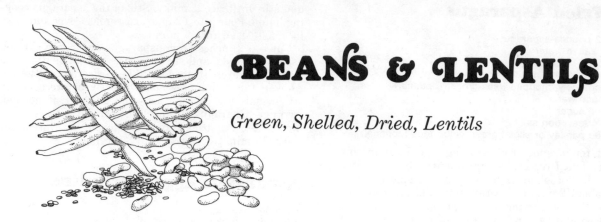

# BEANS & LENTILS

*Green, Shelled, Dried, Lentils*

## GREEN

### Selection and Storage

**Selection:** Choose crisp bean pods that are bright in appearance, free of blemish, and firm.

**Buying:** Allow about 1/4 pound for each serving.

**Storage:** Put unwashed beans in plastic bags in the refrigerator. Use as soon as possible.

### Cutting and Cooking

When ready to use, rinse beans and snip off both ends. In addition to whole or cut green beans, consider for variety thinly slicing or chopping them, cutting them in long slanting slices as Chinese cooks do, or cutting them French-style in long thin slivers (housewares shops sell inexpensive gadgets for cutting beans lengthwise in thin slivers). How you cut them affects their flavor as well as the cooking time.

**Cook green beans** either in a quantity of boiling salted water or in just a little water if beans are cut in small pieces and a heavy pan with tight-fitting lid is used for cooking. Cooking time can vary from as brief as 3 minutes for tender, young beans cut small, to as long as 20 minutes for large, mature beans. Some varieties cook more quickly than others. The broad Italian (or Romano) green beans and Chinese Long beans (called yard-long) can both be cooked in most of the same ways as green or wax beans, but they cook tender in less time.

Fresh beans are best when cooked until just tender but still bright green and slightly crisp. If they reach this point before you're ready to serve, it is best to stop the cooking, drain, if necessary, and rinse quickly in cold water. To reheat, add 1 to 2 tablespoons water to pan and heat quickly. Or reheat them in one of the flavorful sauces suggested below.

### Seasoning and Serving

Fresh green beans can be dressed simply and deliciously with melted butter, salt, and pepper. Add herbs to taste—such as basil leaves, oregano leaves, chives, or thyme leaves. Besides the simple butter sauces that follow, also consider Pesto Sauce or Pesto Butter (recipes pages 6-7).

### Beans Maitre D' Hotel

Cut and cook beans any of the ways suggested above; drain. For each 1 pound beans (hot or cold), heat about 4 tablespoons butter or margarine in a wide pan. Add beans and sauté just until heated through. Sprinkle with about 1 tablespoon minced parsley, a few drops lemon juice, and salt and pepper to taste.

### Green Beans Amandine

Cut and cook 1 pound beans following directions under *cutting and cooking* above. In a frying pan, sauté about 1/4 cup sliced or slivered almonds in 3 tablespoons butter until lightly brown. Either pour over hot cooked beans or add the drained beans to the hot butter and stir until heated. Season with salt and pepper to taste.

---

### Butter-Steamed Green Beans

Remove strings and stem ends from 1 pound green beans and cut in 1-inch lengths. Melt 2 tablespoons butter in a wide electric frying pan set at high heat or in a wide frying pan over direct high heat. Add beans and 5 tablespoons water. Cover and cook, stirring occasionally, for 7 minutes. Salt to taste. Cook frozen, thawed, cut green beans for 5 minutes. Makes 4 to 5 servings.

### Cream-Glazed Green Beans

Prepare green beans as suggested for Butter-Steamed Green Beans, (above). After cooking for 7 minutes, remove cover, salt to taste, and add

5 tablespoons whipping cream. Cook, stirring, until liquid is almost all gone. Serves 4 to 6.

**Green Beans with Sour Cream Egg Sauce**

Prepare green beans following directions for Butter-Steamed Green Beans on page 14. After cooking for 7 minutes, remove covered beans and salt to taste. Add 2 tablespoons whipping cream and cook until liquid is almost gone.

Turn off heat and mix in 1/2 cup sour cream blended with 1/2 teaspoon paprika and 2 chopped, hard-cooked eggs. Makes 5 servings.

# Stir-Fried Green Beans

About 1 pound green beans
Boiling salted water
1 tablespoon salad oil
1 to 3 cloves garlic, minced or mashed
1 tablespoon water
1 teaspoon sugar
½ teaspoon *each* basil leaves and salt
Chopped parsley (optional)

Remove ends from beans and cut slanting slices about 1 1/2 inches long. Drop into a large quantity of boiling water and precook until just tender when pierced with a fork (3 to 5 minutes). Drain, cool quickly with cold water, drain again, and set aside. (This much can be done ahead.) Before starting to cook, prepare all remaining ingredients and have within reach of your range. Heat a 10-inch or larger frying pan or a Chinese wok over high heat, then put in oil. As soon as oil is hot enough to ripple when the pan is tipped, put in garlic; quickly stir and turn with a wide spatula until it starts to brown (about 30 seconds). Put in the beans, water, sugar, basil, and salt. Keep turning with a spatula until heated through (about 1 to 2 minutes). Turn out onto a warm serving dish and garnish with parsley, if desired. Makes about 4 servings.

# Italian Green Beans with Onions

Follow the recipe above for Stir-Fried Green Beans, using either regular or Italian green beans. (Pre-cook Italian green beans only about 2 minutes.) Add 1 large or medium-sized onion, sliced 1/4 inch thick and separated into rings. Put the onions into the hot oil first, stir-fry until browned, then add garlic and proceed as directed. Instead of parsley, sprinkle with about 2 tablespoons shredded Parmesan cheese when you serve.

# Green Beans, Mediterranean Style

1 pound green beans
1 large onion, chopped
4 tablespoons olive oil or salad oil
½ green pepper, seeded and chopped
1 stalk celery, chopped
½ cup water
Salt and pepper to taste

Remove ends and strings from beans. Cut beans in 2-inch lengths.

In a wide frying pan, cook onion in oil over high heat, stirring, until onion is limp but not brown. Add green pepper and celery and cook, covered, over moderate heat for 5 minutes, stirring occasionally.

Add the beans to the cooking vegetables along with water and continue to cook, covered, over moderate heat for 20 minutes, stirring occasionally. Serve hot or warm; do not chill. Season with salt and pepper. Makes about 6 servings.

# Green Beans Oriental

2 packages (9 oz. *each*) French cut green beans
Boiling salted water
2 tablespoons butter or margarine
2 tablespoons minced onions
1 package (7 or 8 oz.) bean sprouts
1 can (12 oz.) water chestnuts, drained and sliced
1 can (10½ oz.) condensed cream mushroom soup
½ soup can milk
½ cup shredded sharp Cheddar cheese
1 can (3½ oz.) French fried onion rings

Cook green beans in boiling salted water as directed on the package. Drain and set aside. Melt butter in a frying pan. Add onion, bean sprouts, and sliced chestnuts; cover pan and cook for 3 to 4 minutes. Put half of the green beans in a buttered 2 1/2-quart casserole; spread with half the bean sprout mixture. Combine the soup with milk; spoon half of it over vegetables. Repeat layers of beans, bean sprouts, and soup. Sprinkle with cheese and bake, uncovered, in a 400° oven for about 25 minutes. Remove from oven and cover with onion rings. Return to oven for about 5 minutes, until hot and bubbly. Makes 8 to 10 servings.

# Green Beans in Swiss Cheese Sauce

1½ pounds green beans
Boiling salted water
¼ cup (⅛ lb.) butter or margarine
¼ cup chopped onion
½ pound mushrooms, sliced
3 tablespoons all-purpose flour
1 teaspoon salt
⅛ teaspoon *each* pepper, thyme leaves, and marjoram leaves
1 cup milk
1 cup shredded Swiss cheese (about 4 oz.)
⅓ cup Sherry or milk

Break off ends and cut beans in about 2-inch lengths. Cook in the boiling water until just tender, drain immediately, and set aside. Meanwhile, heat the butter in a frying pan and sauté the onion and mushrooms until tender (about 6 minutes). Stir in the flour, salt, pepper, thyme, and marjoram. Gradually stir in the milk. Bring to boil, stirring, and cook until thickened.

Remove from heat and blend in 1/2 cup of the cheese and the Sherry. Stir in beans and turn into a shallow 1 1/2-quart casserole. Sprinkle the top with remaining cheese. Bake, uncovered, in a 400° oven until heated through, about 30 minutes (longer if refrigerated). Makes 6 servings.

## Marinated Green Beans

2 pounds green beans
2 quarts water
3 tablespoons coarse (Kosher-style) salt
2 teaspoons *each* mustard seed and dill weed
1 teaspoon *each* crushed small dried hot chile peppers and dill seed
4 cloves garlic
2 cups *each* water and white vinegar
⅔ cup sugar

Snip ends from beans and wash thoroughly; leave whole or cut in half if long. In a large pan bring the 2 quarts water to boil; add 1 tablespoon of the salt and the beans. Return to a boil and cook beans, uncovered, for about 5 minutes or until beans are just tender-crisp. Drain immediately and cool. Pack beans into 4 refrigerator containers (1-pint size). Into each container put 1/2 teaspoon *each* mustard seed and dill weed, 1/4 teaspoon *each* chiles and dill seed, and 1 clove garlic.

Bring to a boil the 2 cups water, vinegar, sugar, and remaining 2 tablespoons salt; pour over beans. Cool, cover, and chill overnight or as long as 2 weeks. Makes 4 pints.

## Lemon Green Beans

1½ pounds green beans, cut in 1½-inch lengths (or two 9 oz. packages frozen, cut green beans)
    Boiling salted water
3 tablespoons melted butter or margarine
½ teaspoon salt
¼ teaspoon pepper
1 tablespoon minced parsley
    Juice of 1 lemon

Cook beans in boiling salted water until tender-crisp (or cook as directed on the package). Drain and place in serving dish. Pour a mixture of the butter, salt, pepper, parsley, and lemon juice over the top. Serve hot. Makes about 6 servings.

## Bean Salad with Basil Dressing

2 packages (9 oz. *each*) frozen French cut green beans, thawed and drained
1 can (about 15 oz.) garbanzos, drained
1 cup thinly sliced celery
1 clove garlic, minced or mashed
¼ cup *each* wine vinegar and salad oil
1 tablespoon sugar
½ teaspoon *each* salt and crushed basil leaves

In a bowl combine the green beans, garbanzos, celery, and garlic. Combine the vinegar, oil, sugar, salt, and

basil; blend well and pour over beans. Mix salad to blend; cover and refrigerate for up to 4 hours. To serve, mix salad gently and place in a shallow serving bowl. Makes 6 to 8 servings.

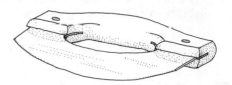

## Tomato Bean Salad

1 package (9 oz.) frozen cut green beans
1 package (10 oz.) frozen lima beans
    Boiling salted water
3 medium-sized tomatoes
1 large green onion, thinly sliced, including part of tops
¼ teaspoon basil leaves
    Dash liquid hot pepper seasoning
¼ cup bottled Italian-style dressing
¼ cup sour cream (or ¼ cup yogurt and ½ teaspoon sugar)
    Tomato wedges for garnish (optional)

Cook green beans and the limas separately in boiling salted water as directed on the packages. As soon as each is tender, rinse quickly in cold water; drain well. Combine the beans and limas in a bowl. Peel the tomatoes, cut in halves, and squeeze gently to remove most of the seed pockets; then cut in cubes. Add tomato, onion, basil, and hot pepper seasoning to the beans. Pour in Italian dressing and mix gently. Cover and refrigerate for 3 to 6 hours. Just before serving, mix in the sour cream and garnish with additional tomato, if desired. Makes 6 servings.

## Baked Green Beans Béchamel

3 packages (9 oz. *each*) frozen French style green beans, thawed
½ cup water
¾ teaspoon salt
½ cup regular strength chicken broth
1 tablespoon cornstarch
2 cups (1 pt.) sour cream
2 teaspoons sugar
⅛ teaspoon ground nutmeg
⅔ cup slivered almonds
1 tablespoon butter or margarine
2 tablespoons shredded Parmesan cheese

Cook beans in the water over highest heat, uncovered, until liquid is boiling, then cook 5 minutes more, stirring frequently; drain well. Mix beans with salt and spread in a shallow 2-quart casserole.

Gradually blend broth with cornstarch, then cook, stirring, until very thick and clear. Blend in sour cream, sugar, and nutmeg, and cook, stirring, just until simmering. Pour hot sauce over beans. Sauté almonds in butter until lightly browned, then spoon decoratively around the side of the casserole. Sprinkle with cheese. Cover and refrigerate up to 24 hours.

Bake in a 375° oven for 40 minutes or until well heated and top is lightly browned. Makes 10 to 12 servings.

# SHELLED

## Selection and Storage

**Selection:** Fresh green limas or cranberry beans may be available shelled or in pods. Or you can buy packages of shelled limas from the freezer case. If you buy them unshelled, choose beans with thick, broad pods that have a glossy sheen and are well filled with large seeds.

**Buying:** Allow 3/4 to 1 pound unshelled beans for each serving.

**Storage:** Keep unshelled beans in the refrigerator. Use as soon as possible.

## Cutting and Cooking

When ready to use, wash pods and cut off the thin outer edge with scissors to facilitate shelling. Slip out the beans, as if shelling peas.

Cook fresh shelled beans in boiling salted water. Boil gently, covered, until tender, about 12 to 20 minutes for limas, about 20 to 25 minutes for cranberry beans; drain.

## Seasoning and Serving

Green lima beans, shelled and cooked following directions given above (or frozen limas cooked as directed on the package), can be enhanced with the addition of melted butter, salt, pepper, or a sprinkling of lemon juice. To these basic seasonings, other more elaborate dressings can be added. A few are included below; each makes enough for 4 pounds fresh lima beans in shells or about 4 to 5 servings.

### Lima Beans Lyonnaise

Chop 2 medium-sized onions and sauté in 1/4 cup melted butter or margarine until tender. Add the cooked and drained beans and heat through. Season with salt and pepper to taste; sprinkle with minced parsley.

### Creamed Lima Beans

Combine the cooked and drained beans in a saucepan with 2 tablespoons melted butter or margarine and 1 tablespoon minced parsley. Beat 2 egg yolks with 3/4 cup half-and-half (light cream); pour over beans. Cook over low heat for 2 to 3 minutes, stirring, until cream is hot and slightly thickened. Do not boil. Serve at once.

### Lima Beans aux Fines Herbes

To cooked and drained beans, add 1/4 cup (1/8 lb.) melted butter or margarine, 1 teaspoon lemon juice, 1 tablespoon *each* minced chives and parsley, and 1 teaspoon tarragon leaves. Cover and simmer until hot. Stir in 1/4 cup finely chopped mushrooms sautéed in butter, if desired.

# Cranberry Bean Minestrone

1½ pounds meaty beef shank (about 3 slices)
  1 *each* carrot, onion, celery stalk, cut-up
  2 teaspoons salt
  ¼ teaspoon pepper
3½ quarts water
  2 large tomatoes, peeled and quartered
  3 small new potatoes, peeled and sliced
  2 cups shelled cranberry beans (about 1¼ lbs. beans)
  ¼ cup pasta stars or other small macaroni
  12 Italian green beans, cut in 1-inch pieces
  2 medium-sized zucchini, sliced
  2 leeks (white part only), sliced
1½ cups shelled peas (about 2 lbs. peas in pods)
  2 teaspoons sugar
  ½ teaspoon basil leaves
    Freshly grated Parmesan cheese (about 1 cup)

In a large soup pot place the beef shanks, carrot, onion, celery, salt, pepper, and water. Bring to a boil and simmer 2 1/2 hours. Let cool slightly, then strain the stock. Cut meat from bones in bite-sized pieces. Discard bones and vegetables.

Add to the strained broth the tomatoes, potatoes, cranberry beans, and pasta; simmer 20 minutes. Add Italian green beans, zucchini, leeks, peas, sugar, basil, and the reserved meat. Simmer 10 minutes longer. Serve in a large soup pot and pass the cheese in a bowl to spoon over. Makes about 8 to 10 servings.

# Cranberry Bean Cassoulet

Cranberry beans are distinguished from other shelled beans by their variegated magenta and cream marblings. However, when heated they lose their brilliant coloring, turning creamy white all over, and are quite mealy in texture, resembling cooked dried beans.

  2 pounds boneless pork loin, cut in 1¼-inch cubes
  2 tablespoons olive oil
  1 medium-sized onion, finely chopped
  2 cups water
  2 cloves garlic, minced
  1 bay leaf
  ½ teaspoon thyme leaves
  1 tablespoon beef stock base or 3 beef bouillon cubes
  1 tablespoon canned tomato paste
  3 whole cloves
  1 whole carrot, peeled
  3 cups shelled cranberry beans
  ½ teaspoon salt
  6 garlic sausages or old-fashioned frankfurters

Using a large frying pan or Dutch oven, brown meat in oil, turning to brown all sides. Remove meat with a slotted spoon to a 3-quart baking dish; sauté onion in drippings remaining in pan. Pour water into pan and scrape up drippings; then pour over meat. Add garlic, bay leaf, thyme, beef stock base, and tomato paste. Poke cloves into carrot and add to the pot, along with cranberry beans and salt. Cover and bake in a 350° oven for 2 hours. Discard carrot.

Arrange sausages on top and continue baking, uncovered, 10 minutes longer. Makes 6 servings.

## Curried Lima Beans

  5 pounds fresh lima beans in the pod, shelled *or* 2
    packages (10 oz. *each*) frozen large lima beans
    Boiling salted water
  ¼ cup (⅛ lb.) butter or margarine
  2 tablespoons *each* finely chopped onion and all-purpose
    flour
  2 teaspoons curry powder
  1½ cups milk
  ½ cup salted round butter cracker crumbs
  4 tablespoons melted butter or margarine

Cook beans in boiling salted water until tender; drain thoroughly and turn into greased casserole (about 1 1/2 quarts). In a saucepan, melt the 1/4 cup butter; add onions and sauté just until soft. Add flour and curry powder, stirring until smooth. Gradually add milk; stir until thickened. Pour sauce over beans; mix gently to coat beans. Toss cracker crumbs with the 4 tablespoons melted butter. Sprinkle over top of beans. Bake in a 400° oven for 15 minutes. Makes 6 servings.

## Baby Lima Bean Soup

  2 packages (10 oz. *each*) frozen baby lima beans, thawed
  2 cans (about 14 oz. *each*) regular strength chicken broth
  1 medium-sized onion, sliced
  1 carrot, peeled and cut in 4 pieces
  1 small stalk celery
  ¼ teaspoon salt
  4 slices bacon, cooked crisp

In a 2 to 3-quart saucepan, combine lima beans, chicken broth, onion, carrot, celery, and salt. Bring to a boil, reduce heat, and simmer until lima beans are tender (about 15 minutes). Whirl soup (about half at a time) in a blender until puréed. Pour the purée through a wire strainer back into the pan and reheat. (For a less smooth soup do not strain.) Serve in a heated soup tureen, if desired. Crumble bacon over. Serves 4.

# DRIED

## Selection and Storage

**Selection:** Dried beans should have a bright, uniform color. Loss of color will mean they have lost freshness. In this state, it will take longer to cook them. Beans should all be the same size; mixed sized will mean uneven cooking. Smaller beans cook faster than larger ones.

Discard any foreign material, cracked seed coats, and pinholes in beans caused by insects.

**Buying:** One pound of beans as purchased usually makes 7 to 9 (3/4 cup) servings.

**Storage:** Keep beans tightly covered and stored in a dry, cool place. Will keep several months.

## Cooking Suggestions

First, wash beans thoroughly discarding any foreign material.

Most recipes call for the dried beans to be soaked before cooking. There are two equally effective methods of soaking beans: the Old-Fashioned method of soaking the beans overnight, and the 2-minute boil and 1 to 2-hour soak method.

For the **Old-Fashioned method,** use 2 1/2 to 3 cups water for each cup beans; soak overnight. Then, if possible, use the soaking water in the recipe to reserve all the natural flavor and nutrients.

For the **2-minute boil and soak method,** bring beans to a boil, using 2 1/2 to 3 cups water for each cup beans; boil for 2 minutes. Remove from heat, let stand 1 to 2 hours, covered.

A cup of dried beans, cooked, makes 2 to 2 1/2-cups. Cooking times vary considerably for different varieties. The following times are approximate but serve as a guide:

  Cranberry beans—about 2 hours
  Great Northern (large white) beans—1 to 1 1/2
    hours
  Kidney beans—about 2 hours
  Large limas—about 1 hour
  Small limas—about 45 minutes
  Navy (small white) beans—1 1/2 to 2 hours
  Pinto (pink) beans—1 1/2 to 2 hours
  Soy beans—about 3 hours

## In-Between Beans

  3 cups dried white beans
  8 cups water
  ¼ pound salt pork, diced
  1½ cups diced onion
  4 tablespoons dark molasses
  2 tablespoons *each* dry mustard, prepared mustard, and
    Worcestershire
  1 cup beer
  3 teaspoons salt
  ¼ cup firmly packed brown sugar
  6 drops liquid hot pepper seasoning, or to taste
  1 cup catsup
  1 large can (15 oz.) tomato sauce

Cover beans with the water and soak overnight (or cover with the water, bring to a boil for 2 minutes, remove from heat, cover, and allow to soak for 1 hour). Without draining beans, place in a large (about 5-quart) baking dish. Add pork, onion, molasses, dry mustard, prepared mustard, Worcestershire, beer, salt, sugar, and liquid hot pepper seasoning. Bake, covered, in a 300° oven for 3 to 4 hours or until beans are almost tender. Remove from oven, stir in catsup and tomato sauce; bake, uncovered, 2 more hours. Serves 12 to 16.

## Garbanzo and Green Olive Salad

1 jar (5 to 6 oz.) or 1 cup salad-style Spanish olives with pimiento, or 1 jar (5 oz.) pimiento-stuffed Spanish olives, chilled
2 cans (about 1 lb. each) garbanzos, chilled
½ cup each chopped onion and celery
2 cloves garlic, minced or mashed
1 tablespoon chopped parsley
½ teaspoon grated lemon peel
3 tablespoons lemon juice
⅛ teaspoon cayenne pepper
⅓ cup olive oil or salad oil
   Iceberg lettuce leaves or romaine leaves
1 cup cherry tomatoes, halved, or 2 large tomatoes, peeled and cut in wedges

Turn olives into a wire strainer and drain thoroughly. Then slice olives and put in a bowl. Drain garbanzos and add to the olives with the onion, celery, garlic, parsley, lemon peel, lemon juice, and cayenne. Drizzle over the oil and mix lightly until all the vegetables are coated with oil. Line a salad bowl with lettuce leaves. Spoon salad over the leaves and garnish with tomatoes. Makes 4 servings.

## Baked Soybeans

1½ cups dried soybeans
   Water
1 small onion, finely chopped
¼ pound salt pork, diced
3 tablespoons molasses
1 cup catsup
1½ teaspoons each salt, dry mustard, and Worcestershire
¼ teaspoon pepper

**To soak soybeans:** Rinse beans, put in a large bowl, and add 3 cups water for each cup of soybeans. Cover and soak 6 to 8 hours or overnight. Drain and reserve liquid for cooking. Pick over beans and discard loose, fibrous skins.

**To cook soybeans:** Pour the beans and soaking water into a large pan, adding more water if needed to cover the beans. Cover pan, bring to a boil, reduce heat, and simmer for about 3 hours or until beans are tender; add more water, if needed, to keep beans from sticking; drain.

Combine cooked beans, onion, salt pork, molasses, catsup, salt, mustard, Worcestershire, and pepper in a 1 1/2-quart casserole. Cover and bake in a 300° oven for 30 minutes. Remove lid, stir beans well, and continue to bake, uncovered, for 30 minutes longer or until sauce is thick. Makes 6 servings.

## Marinated Kidney Beans

1 can (15 oz.) red kidney beans
½ cup finely chopped onion
⅛ teaspoon each salt and sugar
   Dash pepper
½ cup red wine vinegar
   Chopped parsley

Place beans and their liquid in a medium-sized bowl. Mix in onion, salt, sugar, pepper, and vinegar. Cover

and chill for at least 24 hours. Drain well; serve as an appetizer or on a relish tray in lettuce cups, sprinkled with parsley. Makes 4 servings.

## Fava Bean Vegetable Soup

1 cup dried fava beans
4 cans (14 oz. each) regular strength beef broth
1 medium-sized potato, peeled and diced
1 stalk celery, finely chopped
1 large green apple, peeled and sliced
1 large onion, chopped
2 teaspoons curry powder
4 tablespoons (⅛ lb.) butter or margarine
2 teaspoons Worcestershire
1¾ cups milk

Rinse beans and remove any foreign material. In a 6-quart pan, bring beans and beef broth to boiling; boil 2 minutes, then remove from heat, cover, and let stand 1 hour. Add the potato and celery. Return to boiling, then simmer, covered, until tender (about 3 hours).

Sauté apple, onion, and curry in butter until apple and onion are soft; add to beans. Stir in Worcestershire and milk. Bring mixture back to simmer. Whirl in blender, then pour through a fine wire strainer, or press mixture through a wire strainer. Makes 6 servings.

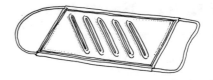

## Chile Bean Casserole

2 cans (15 oz. each) pinto beans, drained
1 can (15 oz.) chile with beans
1 can (8 oz.) bean dip with jalapeños
1 green pepper, seeded and chopped
½ cup minced green onion
2 tablespoons tomato-based chile sauce
1 cup shredded Cheddar cheese

In a shallow casserole, about 2-quart size, combine pinto beans, chile with beans, bean dip with jalapeños, green pepper, green onion, chile sauce; stir to blend, then spread mixture in an even layer in the casserole. Sprinkle with Cheddar cheese. Bake, uncovered, in a 325° oven for 45 minutes or until heated through. Makes 6 to 8 servings.

## Sweet and Sour Beans

6 slices bacon, diced
1 onion, chopped
1 can (1 lb.) Italian style pear-shaped tomatoes
3 tablespoons red wine vinegar
¼ cup firmly packed brown sugar
¼ teaspoon each ground ginger and nutmeg
1 tablespoon soy sauce
2 cans (about 15 oz. each) red kidney beans, drained
1 tablespoon minced parsley

In a large frying pan, cook bacon until crisp; remove bacon with a slotted spoon and drain. Add onion to pan drippings and sauté until limp. Add tomatoes, vinegar, brown sugar, ginger, nutmeg, and soy sauce.

Bring tomato mixture to boiling over highest heat, stirring to break up tomatoes; boil to reduce amount to about 1 1/2 cups. Add drained beans to pan. Cover and simmer until beans are heated. Stir in bacon and, if desired, transfer to a serving dish. Sprinkle with parsley. Makes 6 servings.

## Refried Beans

1 pound dried pinto or pink beans, cleaned
5 cups water
1 or 2 medium-sized onions, diced (optional)
½ to 1 cup hot bacon drippings, butter, or lard
   Salt to taste

Combine beans in a pan with water and onions. Bring to a boil, cover, and remove from heat for 2 hours (or soak beans in cold water overnight). Return to heat, bring to a boil, and simmer slowly until beans are very tender (about 3 hours). Mash beans with a potato masher and add bacon drippings, butter, or lard. Mix well; continue cooking, stirring frequently, until beans are thickened and fat is absorbed. Salt to taste. Serve or reheat. Makes 6 to 8 servings or 5 to 6 cups.

## Olive Baked Navy Beans

2 cups dried navy (small white) beans
  Water
1 can (about 1 lb. 12 oz.) tomatoes, strained
1 small onion, finely minced
1½ cups finely chopped celery
½ cup olive oil
½ cup pimiento-stuffed olives, cut into small pieces

Soak beans in water (cover beans by 1 inch) by either method described under *cooking suggestions*, page 18. Without draining, simmer beans, covered, just until tender and skins split (about 1 hour). Combine in a frying pan tomatoes, onion, celery, and olive oil. Cook over medium heat until celery is tender. Combine beans and olives in casserole. Pour vegetable sauce over beans. Bake, covered, in a 300° oven 3 hours. Makes about 8 servings.

## Hot Bean Salad

1 can (15 oz.) garbanzos
1 can (about 15 oz.) pinto beans
1 can (about 15 oz.) red kidney beans
1 cup thinly sliced celery
1 small red onion, thinly sliced
¼ cup minced parsley
⅓ cup olive oil
¼ cup wine vinegar
½ teaspoon salt
¼ teaspoon *each* pepper and dry mustard
2 tomatoes, cut in wedges

In a saucepan, combine garbanzos, pinto beans, and kidney beans. Bring beans to simmering over medium heat, then simmer gently, until beans are hot.

Meanwhile, combine in a shallow, heat-proof serving bowl the celery, onion, and parsley. In a jar, combine the olive oil, vinegar, salt, pepper, and mustard. Cover

and set aside. Drain beans and add to vegetables. Shake dressing in jar to mix and stir into salad. Arrange tomato wedges around the salad. Makes 4 to 6 servings.

# LENTILS

## Baked Lentils with Cheese

12 ounces (1¾ cups) lentils, rinsed
2 cups water
1 bay leaf
2 teaspoons salt
¼ teaspoon pepper
⅛ teaspoon *each* marjoram leaves, whole sage, and thyme
   leaves, crumbled
2 large onions, chopped
2 cloves garlic, minced or mashed
1 can (1 lb.) tomatoes
2 large carrots, sliced ⅛ inch thick
½ cup thinly sliced celery
1 green pepper, seeded and chopped.
2 tablespoons finely chopped parsley
3 cups (about 10 oz.) shredded sharp Cheddar cheese

In a shallow baking dish (about 9 by 13 inches), mix the lentils, water, bay, salt, pepper, marjoram, sage, thyme, onions, garlic, and tomatoes. Then cover tightly with foil and bake in a 375° oven for 30 minutes.

Uncover; stir in carrots and celery. Bake, covered, for 40 minutes more or until vegetables are tender. Stir in green pepper and parsley. Sprinkle cheese on top; bake, uncovered, for an additional 5 minutes or until cheese melts. Makes 6 servings.

## Italian Sausage and Lentils

2 tablespoons butter or margarine
6 Italian sausages (about 1½ lbs.)
1 large onion, chopped
1 clove garlic, minced or mashed
12 ounces (1¾ cups) lentils, rinsed
2½ cups water
1 whole bay leaf
⅛ teaspoon *each* marjoram leaves, thyme leaves, and
   oregano leaves, crumbled
3 cups turnips cut in ⅓-inch-thick slices
1 can (1 lb.) tomatoes
½ cup dry red wine (or 2 tablespoons wine vinegar with ⅓
   cup water)
½ cup chopped parsley

Melt the butter in a Dutch oven over medium heat and brown the sausages on all sides; remove from pan and reserve. Add onion and garlic to the pan and sauté until onion is soft (about 5 minutes). Cut sausages in half lengthwise and return to pan. Add the lentils, water, bay, marjoram, thyme, and oregano. Cover and simmer for 30 minutes.

At this point add turnips, tomatoes, and wine to Dutch oven; simmer 40 minutes more or until lentils are just tender. Stir in about half the parsley, turn into serving bowl, and sprinkle remaining parsley on top. Makes 6 servings.

# Lentil-Wheat Pilaf

½ cup lentils, rinsed
2 cups water
1 teaspoon salt
3 tablespoons olive oil
1 small onion, chopped
¼ pound mushrooms, sliced
½ cup quick-cooking cracked wheat
   Unflavored yogurt
   Sliced green onions, including part of the tops

Combine the lentils, water, and salt in a saucepan. Bring to a boil, cover and simmer for 20 minutes; set aside.

   Heat the oil in a 10-inch frying pan over medium heat; sauté onion, mushrooms, and wheat until onion is soft (about 5 minutes). Pour lentils and water over wheat, bring to a boil, cover, and simmer for 15 minutes or until wheat is just tender. Serve with yogurt and green onions to spoon on top. Makes 4 servings.

# Spicy Baked Lentils

1 package (12 oz.) lentils
1 teaspoon salt
   About 3 cups water
4 slices bacon, finely diced
2 cans (8 oz. *each*) tomato sauce
1 medium-sized onion, finely chopped
¼ cup firmly packed brown sugar
2 tablespoons prepared mustard
⅓ cup molasses

Sort through lentils, rinse, and drain. Place lentils in a casserole with a lid (2 1/2 or 3-qt. size). Add the salt, 3 cups of the water, bacon (uncooked), tomato sauce, onion, brown sugar, mustard, and molasses. Stir well.

   Cover dish and bake in a 350° oven until lentils mash readily and liquid is bubbly and thickened (about 2 hours). Stir gently about every 30 minutes, adding a little more water if sauce becomes too thick. Makes about 8 servings.

# BEAN SPROUTS

## Selection and Storage

**Selection:** Choose crisp bean sprouts with beans attached.

**Buying:** Allow about 1/4 pound for each serving; 1 pound makes 4 to 5 servings.

**Storage:** Store, unwashed, in plastic bag in refrigerator. Use as soon as possible.

## Cooking Suggestions

Rinse bean sprouts in cool water just before using them and sort out any that are discolored; drain. You may also want to pinch off the root ends, especially any that are shrivelled and brown. (For some Oriental dishes, both ends of the sprouts are removed.) You can use the sprouts raw or blanched in salads and sandwiches (blanching removes the slight astringency that bean sprouts have).

**To blanch bean sprouts,** plunge them into a quantity of boiling salted water; boil about 2 minutes. Immediately drain and rinse in cold water to cool quickly; drain again.

   Bean sprouts add a pleasing crunchiness to Oriental stir-fry dishes or when cooked with other vegetables. They cook tender-crisp in about 2 minutes. After that they will rapidly lose their crunchiness if allowed to cook much longer.

# Stir-Fried Bean Sprouts

3 to 4 cups (about ¾ lb.) fresh bean sprouts
1 tablespoon salad oil
1 to 2 teaspoons finely chopped fresh ginger (optional)
1 to 2 cloves garlic, minced
1 tablespoon water
1 teaspoon sugar
½ teaspoon salt
   About 2 tablespoons chopped parsley or sliced green onions, including part of tops

Rinse bean sprouts and allow to drain while you prepare and measure all remaining ingredients; set them within reach of your range. Heat a large frying pan or wok over high heat, then put in oil. As soon as oil is hot enough to ripple when the pan is tipped, add ginger (if used) and garlic; quickly stir and turn with a wide spatula until browning starts (about 30 seconds). Put in sprouts, 1 tablespoon water, sugar, and salt. Keep turning with spatula until heated through (about 2 minutes). Or after adding bean sprouts to the pan, you can put a cover on and cook, uncovering often to turn vegetables with a spatula, just until tender (about 1 to 2 minutes). Turn out onto a warm serving dish and garnish with the chopped parsley or sliced green onion. Makes 4 to 5 servings.

## Bean Sprout Sauté

1 tablespoon salad oil
½ cup chopped green onions, including part of tops
1 cup shredded carrots (about 2 carrots)
4 tablespoons water
4 cups (about ¾ lb.) bean sprouts, rinsed
  Salt to taste

Heat oil in a wide frying pan or electric frying pan over highest heat. Add onions, carrots, and water. Cook, stirring, until carrots are slightly tender. Then mix in the bean sprouts and continue to cook, uncovered, over high heat, stirring, until carrots are tender-crisp and bean sprouts are just hot. Season with salt. Serves 4.

## Tuna Salad with Bean Sprouts

1 can (6½ or 7 oz.) tuna, flaked
¾ cup fresh bean sprouts, rinsed
3 teaspoons capers
¼ cup mayonnaise
1 teaspoon soy sauce
  Lettuce leaves
2 whole sweet midget pickles
  Mild, pickled whole sweet cherry peppers

Gently stir together the tuna, bean sprouts, capers, mayonnaise and soy sauce.

Line two salad plates with lettuce leaves. Mound tuna mixture on lettuce. With a sharp knife, slice the pickles lengthwise almost to one end, each in about 4 slices; fan slices out. Use these with the cherry peppers as garnish for the salads. Makes 2 luncheon salads.

## Rice and Bean Sprouts

3 tablespoons crushed toasted sesame seed (directions follow)
2 green onions, including part of tops, minced
1 clove garlic, minced or mashed
1 tablespoon salad oil or sesame oil
1½ cups fresh bean sprouts
2 cups hot cooked rice
2 tablespoons soy sauce

Combine sesame seed with onions and garlic and sauté in oil for 3 minutes. Add the bean sprouts and sauté until thoroughly hot. (It may be necessary to add a few drops of water to keep ingredients from sticking.) Add the hot rice and soy sauce and gently mix, being careful not to mash rice grains. Makes 4 to 6 servings.

**Crushed toasted sesame seed.** Place sesame seed in a heavy frying pan. Stirring, cook over medium heat 10 minutes or until golden brown. Turn seed into a mortar, add dash salt, and crush with a pestle or whirl in an electric blender.

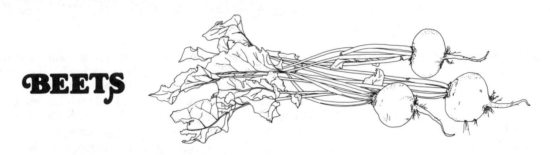

# BEETS

### Selection and Storage

**Selection:** Choose beets that are small with smooth skins, firm slender roots, and a deep red color. Tops should be fresh and green.

**Buying:** 1 pound (a medium-sized bunch) makes 2 to 3 servings.

**Storage:** Cut off top greens about 1 inch above beet crown; leave root ends intact. Put beets in plastic bag and store in refrigerator. Will stay fresh for several days. Save the beet tops if they are fresh and green; store in the refrigerator and use as soon as possible. (Recipes for cooking beet greens as a separate vegetable are on page 52.)

### Cutting and Cooking

To preserve their rich, red color, beets are usually cooked in their jackets. Cooked beets, when peeled, may be cut and seasoned in a variety of ways. Raw beets may also be cooked quickly in

butter and a small amount of water if they are cut in small pieces (see butter-steaming, page 23).

**To cook whole beets,** scrub roots with a vegetable brush. Leave on skin, the rootlet, and about 1 inch of the tops. In a covered pan, cook beets in boiling, salted water to cover until tender; it may take 20 to 60 minutes, depending on size and maturity of beets. Drain and cool quickly under running water. (At this point, you can store beets in the refrigerator to use later.) Rub off skins under running water and use for any recipe that calls for cooked beets.

### Seasoning and Serving

Once the whole beets have been cooked, cooled, and peeled, they may be sliced, diced, shredded, or cut in quarters or julienne strips; tiny, baby beets may be left whole. Season simply with butter, salt and pepper, and a squeeze of lemon or a little wine vinegar, if desired. Or you might try some of the easily prepared seasoning suggestions that follow.

## Beets Beurre Noir

Prepare 1 bunch (1 lb.) beets and cook whole, following directions under *cutting and cooking*, page 22. Peel, slice, and return beets to saucepan. Sauté in 1/4 cup butter or margarine to warm; remove to a hot serving dish. Add another 2 tablespoons butter to the saucepan and brown; add 1 tablespoon lemon juice, 1 tablespoon minced parsley, salt and pepper to taste. Pour over beets and serve.

## Beets with Chive Sauce

Prepare 1 bunch (1 lb.) beets and cook whole, following directions under *cutting and cooking*, page 22. Drain (reserving liquid), peel, and slice beets. In a pan melt 2 tablespoons butter, add 2 tablespoons all-purpose flour, and cook until bubbly. Gradually stir in 1 cup water (from cooking beets) or regular strength chicken broth; cook, stirring, until it boils and thickens. Add 1 tablespoon wine vinegar, 3 tablespoons minced chives or green onion, 1 tablespoon minced parsley, and the sliced beets. Cook until heated through.

## Beets Tarragon

Follow directions for Beets with Chive Sauce above, omitting chives and parsley. Instead add to the sauce 1 teaspoon crumbled tarragon leaves and 1 teaspoon prepared mustard.

---

## Butter-Steamed Beets

Cut tops and root ends from about 2 bunches raw beets. Peel and cut in thin slices (you should have about 4 cups).

Melt 2 tablespoons butter in a wide frying pan over high heat or an electric frying pan on highest heat. Add the sliced raw beets and 6 tablespoons water. Cover and cook, stirring occasionally, for 5 to 6 minutes or until beets are tender; add 2 to 3 more tablespoons water if needed. Season with salt and pepper. Makes 4 to 6 servings.

## Butter-Steamed Beets with Orange

Prepare beets following the directions for Butter-Steamed Beets above, adding 2 tablespoons orange juice and 1/4 teaspoon grated orange peel during the last 2 minutes of cooking time. Pour the beets into serving bowl and blend in 1/2 cup sour cream; salt and pepper to taste. Makes 4 to 6 servings.

## Beets with Beet Greens

Prepare beets as directed for Butter-Steamed Beets above. After discarding coarse stems, wash tender beet leaves; stack leaves and slice crosswise into 1/4-inch strips. Add greens to the beets during the last 2 or 3 minutes of cooking.

# Beet Preserves

```
6 pounds beets, peeled
  Water
8 cups sugar
  Juice and grated peel of 4 large lemons
¼ pound fresh ginger, scraped and sliced
½ pound shelled filberts or almonds, chopped
```

Put raw beets through a food chopper with a coarse blade. Cover with water in a saucepan and cook gently until tender (about 15 minutes). Add sugar, lemon juice, grated lemon peel, and ginger. Cook slowly about 1 hour until beets are translucent; add filberts during last 10 minutes. Pour boiling hot beets into hot sterilized jars, leaving 1/8 inch head space. Adjust caps to seal. Makes about 6 pints preserves.

# Beets with Mustard Butter

```
3 cups diced cooked beets
¼ cup soft butter or margarine
1 tablespoon prepared mustard
1 tablespoon tarragon vinegar
  Water
```

Cook beets whole, following directions under *cutting and cooking*, page 22; drain, peel, and dice. Beat together the butter, mustard, and tarragon vinegar.

Return beets to the pan and reheat quickly in about 1 tablespoon water; turn into a warm serving bowl. Add the mustard butter in small pieces. Mix lightly before serving. Makes 4 to 6 servings.

# Molded Beet Salad

```
2 envelopes unflavored gelatin
½ cup cold water
1 can (16 oz.) diced beets
¼ cup white wine vinegar
3 tablespoons lemon juice
⅓ cup sugar
1 teaspoon salt
  Dash black pepper
1 cup each chopped celery and peeled, chopped
  cucumber
¼ cup chopped green onions, including part of tops
```

Soften the gelatin in the cold water. Combine beets and liquid in can with vinegar, lemon juice, sugar, salt, and pepper, and whirl in the blender just until smooth. Pour into a pan and heat to boiling.

Stir the softened gelatin into the hot beet mixture and continue stirring until dissolved. Chill until partially set. Then stir in the chopped celery, cucumber, and green onion. Pour into a 6-cup mold and chill until set. Serve on a bed of chopped cabbage with a simple mayonnaise or sour cream dressing. Garnish with cucumber slices, if you wish. Makes about 6 servings.

## Beet Borscht

5 to 6 large beets with leafy tops
1 large onion
10 cups water
4 tablespoons seasoned chicken stock base (or 10 chicken bouillon cubes)
1 large potato
8 to 10 tablespoons lemon juice
1 to 1½ tablespoons sugar
   Salt
   Sour cream

Trim the fresh, tender looking leaves from beet stems, discarding stems and coarse leaves. Wash tender leaves and drain well; then chop. Peel raw beets and shred coarsely (you should have 5 to 6 cups). Peel and shred the onion, then combine with beet tops and beets in a large kettle; add water and chicken stock base and bring to a boil. Cover and simmer soup for 15 minutes. Peel the potato and cut in 1/2-inch cubes; add to soup and cook another 15 minutes. Add 6 tablespoons of the lemon juice and 1 tablespoon of the sugar; chill soup thoroughly. Taste cold soup and add salt as needed and as much more of the lemon juice and sugar as you need to achieve the sweet-sour flavor balance you prefer. Cover and chill for as long as a week. Ladle into bowls and add sour cream as desired. Makes about 3 quarts.

## Spiced Pickled Beets

2 cans (16 oz. *each*) small whole beets
1 medium-sized red or white onion, thinly sliced
½ cup white wine vinegar
¼ cup sugar
12 whole cloves
½ teaspoon salt
2 hard-cooked eggs, sliced (optional)

Drain beets, reserving 1 cup of the liquid. In a bowl combine beets and sliced onion. In a saucepan combine the cup of beet liquid, vinegar, sugar, cloves, and salt; bring to a boil, stirring often; pour over beets and onions. Let cool, cover, and refrigerate for 1 or 2 days.

To serve, lift beets and onions from marinade with a slotted spoon. Arrange in a bowl and garnish with eggs, if desired. Makes about 6 servings.

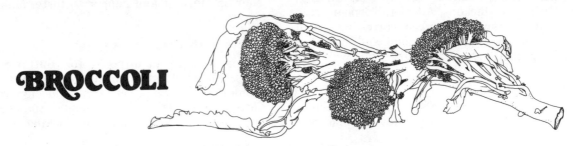

# BROCCOLI

### Selection and Storage

**Selection:** Broccoli heads should have compact clusters of tightly closed dark-green flowerets. Avoid heads with yellowing buds. Fresh, clean stalks should be firm and tender but not too thick.

**Buying:** Allow 1/3 to 1/2 pound for each serving.

**Storage:** Put unwashed broccoli in a plastic bag, seal, and refrigerate. Use within 4 to 5 days.

### Cutting and Cooking

Wash broccoli thoroughly when ready to use, then trim off just the end of the stem. If outer layer of remaining stalk is tough, peel it; then the entire stalk is edible.

**For whole stalks,** if any are more than 1 inch in diameter, make 4 to 6 lengthwise slashes through stems almost to flowerets. This allows the stalk to cook as quickly as the flowerets. Place in a saucepan of about 1-inch boiling salted water or in a steamer. Cover and cook quickly until just tender-crisp (7 to 12 minutes); drain.

**For broccoli flowers and stems,** cut off flowers, leaving some stem on each. Slice peeled stems crosswise about 1/4 inch thick. Slice large flowerets lengthwise through stems to make pieces about 1/4 inch thick. Then flowers and stems cook tender in same time; cook together or separately to serve as two different vegetables. Cook quickly in a small amount of boiling salted water (or steam) until just tender-crisp (about 4 to 6 minutes); drain.

**For thinly sliced or diced broccoli,** cut off flowerets close to the flower buds (they will fall apart); then thinly slice or dice the peeled stems. Cook in a small amount of boiling salted water until just tender (about 4 to 6 minutes); drain.

### Seasoning and Serving

Serve hot cooked broccoli (prepared any of the ways suggested above) seasoned with melted butter, salt and pepper, or use a favorite seasoned butter such as Pesto Butter (recipe page 7). Cold cooked broccoli (whole spears or flowers and stems) are delicious served with Green Herb Sauce or Pesto Mayonnaise (recipes on pages 6-7).

**When you want to cook broccoli ahead,** cook whole stalks or flowers and stems as directed, taking special care not to overcook. Drain at once and immerse in a large quantity of cold water to cool quickly; then drain well. To serve, reheat quickly in a butter or oil-based sauce, or put vegetable in pan with 1 or 2 tablespoons water to reheat quickly.

The following seasoning suggestions are based on about 1 1/2 to 2 pounds broccoli, the amount needed to serve 4 to 6 servings.

### Broccoli Polonaise

Cook whole stalks or broccoli flowers and stems following directions on page 24; drain; keep hot. Brown 1/2 cup fine, dry bread crumbs in 6 table-spoons butter; remove from heat and add 1 chopped hard-cooked egg and 1 tablespoon minced parsley. Sprinkle over hot broccoli.

### Broccoli Mornay

Cook whole stalks of broccoli following *cutting and cooking*, page 24; drain and arrange on a broil-and-serve platter. Cover with about 2 cups Mornay Sauce (recipe page 6) and sprinkle with 3 tablespoons grated Parmesan or Romano cheese; brown quickly under broiler.

### Broccoli with Eggs

Prepare and cook broccoli following *cutting and cooking*, page 24 for thinly sliced or diced broccoli; drain. Arrange in a shallow baking dish; season to taste with salt and pepper; keep warm. Meanwhile, fry 6 eggs in 1/4 cup olive oil or butter; place on the broccoli. Drizzle remaining oil over top and sprinkle with 1/2 cup grated Parmesan or Romano cheese. Brown quickly under broiler.

### Cold Broccoli with Cashews

Cook whole stalks of broccoli following *cutting and cooking*, page 24; drain and place in a serving dish. Combine 1/4 cup olive oil, 2 tablespoons lemon juice, 1/4 teaspoon salt, a dash pepper, and 1 teaspoon chervil; mix well. Drizzle over broccoli; chill. Just before serving, sprinkle with about 1/4 cup salted cashews, coarsely broken.

---

### Butter-Steamed Broccoli

Trim ends and peel tough stems from 1 1/2 pounds broccoli. Cut off flowerets close to blossoms and set aside. Slice the stems thinly. Melt 2 tablespoons butter in an electric frying pan set at high heat or use a wide frying pan over direct high heat. Add broccoli stems and 4 tablespoons water. Cover and cook, stirring occasionally, for 4 minutes. Stir in broccoli blossoms and 3 more tablespoons water. Cover and cook 3 minutes longer, stirring frequently. Season with salt and serve. Makes 4 to 5 servings.

### Broccoli with Cream Cheese Sauce

Prepare broccoli as for butter-steaming. After the salt, add 1 package (3 oz.) cream cheese, diced; stir until melted. Sprinkle with 1 tablespoon lemon juice and serve. Makes 4 to 5 servings.

## Stir-Fried Broccoli

3 to 4 cups cut broccoli (cut flowers and stems as
    directed under *cutting and cooking*, page 24)
Boiling salted water
About 1 tablespoon salad oil
1 to 2 teaspoons finely chopped fresh ginger (optional)
1 or 2 cloves garlic, minced or mashed
1 tablespoon water
1 teaspoon sugar
½ teaspoon salt
    Toasted sesame seed or toasted chopped nuts (almonds,
      cashews, filberts) for garnish (optional)

To precook broccoli, drop into a large quantity of rapidly boiling salted water and cook until just tender when pierced with a fork (3 or 4 minutes). Drain broccoli and cool quickly with cold water, drain it again and set aside.

Before starting to cook, prepare all remaining ingredients and have them within reach of your range. Heat a 10-inch or larger frying pan or wok over high heat, then put in oil. As soon as oil is hot enough to ripple when the pan is tipped, add ginger (if used) and garlic; quickly stir with a spatula until browning starts (about 30 seconds). Put in the broccoli, water, sugar, and salt. Keep turning with a spatula until heated through (about 1 to 2 minutes). Turn out onto a warm serving dish and garnish with sesame seed or nuts, if desired. Makes about 4 to 5 servings.

## Broccoli with Fresh Mushrooms

Trim up to 1/2 pounds mushrooms and cut in 1/4-inch slices. Follow recipe above for Stir-Fried Broccoli, increasing salad oil to 2 tablespoons and omitting ginger and garlic. Add the mushrooms to the hot oil first; stir and fry them until browned. Then add 2 shallots or green onions, chopped, and a dash ground nutmeg; stir and fry about 30 seconds. Then add broccoli and remaining ingredients. Finish as directed.

## Broccoli Stems with Peas

2 cups broccoli stems
1 teaspoon salad oil, butter, or margarine
½ cup sliced green onions, including part of tops
1 package (10 oz.) frozen peas
½ cup water
½ teaspoon salt
½ teaspoon lemon juice
2 teaspoons chopped parsley

Cut broccoli stems in 1/4-inch slices. Heat oil in a wide frying pan over medium-high heat. Stir in broccoli, onions, frozen peas, water, and salt. Cover and cook until vegetables are just tender (about 5 minutes). Mix in lemon juice and chopped parsley. Makes 4 servings.

## Broccoli Lorraine

1½ pounds broccoli
3 slices bacon, cooked crisp, and crumbled
¾ teaspoon salt
⅛ teaspoon *each* pepper and ground nutmeg
½ teaspoon dry mustard
4 eggs
1½ cups half-and-half (light cream)
3 tablespoons freshly shredded Parmesan cheese

Trim and cut broccoli as directed for flowers and stems under *cutting and cooking*, page 24; drain well and turn into a 2-quart shallow baking dish; sprinkle with bacon.

In a bowl, combine the salt, pepper, nutmeg, and mustard. Add eggs and beat lightly with a fork, then stir in half-and-half and Parmesan and pour over the broccoli.

Set baking dish inside a pan of hot tap water so that water comes to within 1/2 inch of dish rim. Bake in a 350° oven for 25 to 30 minutes or until you can shake the pan back and forth and only a 3-inch circle in the center moves. Serve at once. Makes about 6 servings.

## Broccoli Casserole

1 package (10 oz.) frozen chopped broccoli or about 1½ pounds fresh broccoli
2 tablespoons *each* butter and all-purpose flour
¼ teaspoon salt
Dash pepper
1 cup milk
1 tablespoon grated onion
¾ cup mayonnaise
3 eggs, well beaten

Cook frozen broccoli following package directions; drain. Or trim and cook fresh broccoli following direc-

tions for thinly sliced or diced broccoli under *cutting and cooking*, page 24; drain well.

Melt butter in a saucepan and blend in flour, salt, and pepper; cook until bubbly. Gradually stir in milk, and cook, stirring, until thickened and smooth. Remove from heat and stir in onion, mayonnaise, and beaten eggs. Carefully mix in broccoli. Turn into a 2-quart greased casserole and place, uncovered, in a pan of hot water. Bake in a 350° oven about 30 minutes or until custard has set. Makes 6 servings.

## Anchovy-Broccoli Salad

1 package (10 oz.) frozen broccoli spears or about 1 pound fresh broccoli
1½ tablespoons *each* anchovy paste and finely minced green onion, including part of tops
½ clove garlic, minced or mashed
3 tablespoons finely chopped parsley
1 tablespoon chopped pimiento
2 tablespoons vinegar
2 teaspoons lemon juice
¼ cup sour cream or yogurt
½ cup mayonnaise
Dash pepper
1 head iceberg lettuce, shredded
2 medium-sized tomatoes, cut in wedges

Cook broccoli as directed on package or as for whole fresh stalks page 24; drain, cover, and chill.

In a small bowl, or blender bowl, combine anchovy paste, onion, garlic, parsley, pimiento, vinegar, lemon juice, sour cream or yogurt, mayonnaise, and pepper. Stir or whirl until blended.

Mound lettuce on 4 salad plates. Arrange broccoli on top; garnish with tomato wedges and spoon dressing over salads. Pass additional dressing. Serves 4.

# BRUSSELS SPROUTS

## Selection and Storage

**Selection:** Brussels sprouts—these miniature heads of cabbage—should be compact and bright green in color and free from blemishes. Avoid wilted heads or those with yellowish leaves.

**Buying:** Allow 1 pound for 3 to 4 servings.

**Storage:** Remove any loose or discolored leaves and place, unwashed, in a plastic bag. Keep in refrigerator and use within 1 to 2 days.

## Cutting and Cooking

Wash thoroughly and trim off stem ends from Brussels sprouts. Then either cut an "x" into the

stem ends or slice each sprout in half lengthwise. (If there is evidence of insect damage, soak sprouts in lightly salted cold water for about 20 minutes; drain.) Cook, uncovered, in a large quantity of boiling salted water until just tender-crisp, about 7 to 10 minutes. Or cook sprouts in about 1 inch boiling salted water, leaving pan uncovered, for the first 4 to 5 minutes, then cover and cook until just tender, 3 to 5 minutes more; drain.

## Seasoning and Serving

Hot cooked Brussels sprouts can be seasoned simply with melted butter, salt, and pepper. Or flavor the melted butter first with a pinch of basil leaves, dill weed, thyme leaves, or curry powder to taste. You might also sprinkle the sprouts with minced parsley or chives before serving.

The following seasoning suggestions are based on 1 to 1 1/2 pounds Brussels sprouts, trimmed and cooked as directed above—enough for 4 to 6 servings.

**If you want to do the cooking ahead,** be sure to cool the cooked sprouts quickly in cold water to prevent overcooking.

### Brussels Sprouts with Onion Butter

In a frying pan, sauté 1 medium-sized onion, sliced, in 3 tablespoons butter or margarine, until the onion is golden brown. Stir in cooked, drained sprouts (hot or cold), and salt and pepper to taste. Stir and cook just until sprouts are coated with butter and heated through. Sprinkle with chopped parsley or grated Parmesan cheese, if desired, and serve.

### Deviled Brussels Sprouts

In a saucepan, melt 3 tablespoons butter or margarine. Stir in 1 teaspoon prepared mustard, 1/2 teaspoon Worcestershire, and 1/2 tablespoon chile sauce or catsup. Drizzle over hot, cooked Brussels sprouts. Or stir hot or cold Brussels sprouts into the deviled butter and heat, stirring, until blended and hot through. Season to taste with salt and pepper and serve.

## Stir-Fried Brussels Sprouts

3 or 4 cups Brussels sprouts, trimmed and cut in half
    lengthwise
Boiling salted water
1 tablespoon salad oil
1 to 2 teaspoons finely chopped fresh ginger (optional)
1 or 2 cloves garlic, minced or mashed
1 tablespoon water
1 teaspoon sugar
½ teaspoon salt
    Chopped parsley, sliced green onion tops, or toasted
    sesame seed for garnish

Pre-cook sprouts in a large quantity of rapidly boiling salted water until just tender when pierced (2 to 3 minutes). Drain, cool quickly with cold water, drain again, and set aside.

Before starting to cook, prepare all other ingredients and have them within reach of your range. Heat a 10-inch or larger frying pan or wok over high heat, then put in oil. As soon as oil is hot enough to ripple when the pan is tipped, add ginger (if used) and garlic; quickly stir with a spatula until browning starts, about 30 seconds.

Put in sprouts, water, sugar, and salt. Keep turning with a spatula until heated through (about 1 to 2 minutes). Turn out onto a warm serving dish and sprinkle with parsley or another garnish. Makes about 4 to 5 servings.

## Brussels Sprouts with Grapes

2 pounds Brussels sprouts, cooked and hot
2 tablespoons melted butter
3 tablespoons half-and-half (light cream)
½ teaspoon *each* salt and ground nutmeg
1 teaspoon Worcestershire
    Dash pepper
2 cups seedless grapes
    Buttered, toasted crumbs

Coat hot cooked Brussels sprouts with melted butter by shaking together in frying pan. Stir in half-and-half, salt, nutmeg, Worcestershire, dash pepper, and seedless grapes. Heat through. Pour into a serving dish and sprinkle with buttered toasted crumbs. Makes about 6 servings.

## Marinated Brussels Sprouts

1¼ pounds Brussels sprouts or 2 packages (10 oz. *each*)
    frozen Brussels sprouts
¼ cup olive oil or salad oil
2 teaspoons chervil or parsley flakes, crushed
¼ teaspoon garlic powder
½ teaspoon sugar
¼ teaspoon *each* salt and tarragon leaves
⅛ teaspoon pepper
1 teaspoon grated lemon peel
3 tablespoons lemon juice
2 tablespoons white wine vinegar

Trim fresh Brussels sprouts, cut each in half lengthwise, and wash well. Cook as directed under *cutting and cooking*, page 26; drain well. (Or cook the frozen sprouts as directed on the package, drain well, cool, then cut each in half.) Put sprouts into a deep bowl. Combine in a small bowl the oil, chervil, garlic, sugar, salt, tarragon, pepper, and lemon peel. Mix well and pour over sprouts. Cover and refrigerate for at least 2 hours, or overnight; stir several times. To serve, drizzle over them the lemon juice and vinegar; mix well. Serve in a shallow bowl. Makes about 4 to 6 servings.

## Brussels Sprouts Appetizer

2 pounds Brussels sprouts
1½ cups mayonnaise
4 tablespoons *each* sweet pickle relish and capers and
    liquid
1 teaspoon sugar
1 tablespoon grated onion
    Salt and pepper to taste
    Dash cayenne
    Vinegar

Trim and cook Brussels sprouts as directed under *cutting and cooking*, page 26. Combine mayonnaise, sweet pickle relish, capers and liquid, sugar, onion, salt and pepper, cayenne, and enough vinegar to slightly thin the sauce.

Skewer each sprout on a toothpick and serve with a bowl of the sauce to be used as a dip. Or toss sprouts with sauce and serve as salad. Makes about 30 medium-sized appetizers.

# CABBAGE

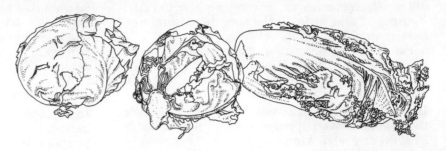

*Green, Red, Chinese*

## Selection and Storage

**Selection:** Choose cabbage heads that are firm and heavy for their size. Outer leaves should look fresh, be a good green or red (depending on variety), and be reasonably free of blemishes. Early in the season, regular green or red cabbage may be less firm and feel somewhat soft. Savoy cabbage, with crinkly leaves, and celery cabbage (also called nappa or Chinese cabbage) have softer, more loosely formed heads. When selecting these cabbages, look for fresh, crisp leaves that are free of blemishes.

**Buying:** Allow 1 pound for 3 to 4 servings.

**Storage:** Place, unwashed, in a plastic bag, close, and refrigerate. Firm, hard cabbage will keep a week or longer. Use soft-head cabbages within a few days.

## Cutting Suggestions

Rinse cabbage head and discard any wilted outer leaves. Shredding is the most versatile way to cut green or red cabbage when you plan to cook it quickly to serve as a hot vegetable. Use a large French knife and slice across the head into fine or wide shreds or use a vegetable shredder. Green cabbage may also be cut in wedges before cooking. Savoy cabbage may be cut in wedges or cooked whole to be cut into serving-size wedges at the table (see recipe below). Celery cabbage is often sliced across the head into about 1-inch-wide pieces.

## Cooking Suggestions For Green Cabbage

**To cook shredded green cabbage,** place in a small amount of boiling salted water; cover and cook rapidly until tender-crisp (3 to 5 minutes), depending on how finely shredded; drain.

**To butter-steam green cabbage,** melt 2 tablespoons butter in a wide frying pan set on high heat (or use a wide frying pan over direct high heat). Add about 5 cups finely sliced cabbage (1 small head) and 4 tablespoons water. Cover and cook, stirring occasionally, for 3 to 4 minutes. Salt to taste and serve.

**To pre-cook green cabbage,** drop cut pieces into a large quantity of rapidly boiling, salted water. Cook until almost tender (1 to 2 minutes), then cool quickly in a quantity of cold water; drain well. Reheat as directed in recipe.

## Seasoning and Serving Green Cabbage

Lightly cooked green cabbage—either boiled or butter-steamed—lends itself to a variety of interesting seasonings. The ideas that follow are based on about 1 pound cabbage, the amount needed for 3 to 4 servings.

### Caraway Cabbage

Follow directions above for cooking shredded green cabbage; drain well. Add 1 cup sour cream, 1 teaspoon slightly crushed caraway seed, and salt and pepper to taste. Stir over low heat until heated through.

### Dilled Cabbage

Follow directions for butter-steaming green cabbage under *cooking suggestions* above. Add 1/4 teaspoon dill weed along with the cabbage. Just before serving, stir in 1/4 cup mayonnaise. Sprinkle with additional dill weed and serve.

### Cabbage Catalina

Follow directions for cooking shredded green cabbage under *cooking suggestions* above; drain. In another pan, sauté 2 tablespoons *each* chopped green pepper and green onions (including part of tops), in 1 tablespoon butter. Stir in the drained cabbage, 2 tablespoons whipping cream, and salt to taste. Reheat quickly and serve.

### Whole Savoy Cabbage with Mornay Sauce

Wash whole cabbage and strip off the outer dark leaves. Remove the core; tie cabbage in cheesecloth and plunge, core end down, into a deep pot of boiling salted water (keep the cabbage submerged by placing a heavy pan on top of cabbage); let cook 3 to 4 minutes. Drain, remove the cheesecloth, and serve hot wedges with Mornay Sauce (recipe page 6).

## Hot Slaw

1 medium-sized head green cabbage, shredded
¼ cup chopped onion
½ cup *each* sour cream and mayonnaise
1 teaspoon *each* prepared mustard and lemon juice
  Pinch of sugar
  Salt to taste

Cook cabbage following directions for *cooking shredded cabbage*, page 28; drain. To the cabbage, add onion, sour cream, mayonnaise, mustard, and lemon juice. Blend in sugar and salt. Stir over low heat until heated through. Makes 6 to 8 servings.

## Curried Cabbage

1 medium-sized head cabbage (about 2 lbs.)
4 tablespoons butter or margarine
1 teaspoon *each* curry powder and sugar
1 clove garlic, minced or mashed
6 slices bacon, cooked crisp and crumbled
  Salt

Core cabbage and cut in 1-inch-wide wedges, then cut each wedge in 1-inch lengths. Pre-cook following directions under *cooking suggestions*, page 28; drain well.

At serving time, melt butter in a 10-inch frying pan; stir in curry powder, sugar, and garlic. Add cabbage and cook, over high heat, stirring, for about 5 minutes. Add bacon and salt to taste. Makes 4 to 6 servings.

## Toasted Hungarian Cabbage

1 large head (about 3 lbs.) cabbage, finely chopped
1 tablespoon salt
½ cup (¼ lb.) butter or margarine
3 tablespoons sugar
  About ¼ teaspoon fresh ground pepper

Mix cabbage with salt and let stand at room temperature for at least 30 minutes. Place in a colander and rinse under cold water, then squeeze out liquid.

Melt butter in a wide frying pan and add cabbage. Cook, uncovered, over medium-low heat for about 30 minutes, stirring frequently. When cabbage becomes limp and turns a brighter color, add sugar and pepper. Continue cooking until cabbage takes on an amber color and a few particles begin to brown lightly and look crisp; stir frequently and gently.

Serve cabbage hot. If made ahead, cover and chill; reheat in a frying pan over moderate heat. Makes about 3 1/2 cups or 6 to 8 servings.

## Cabbage Plus

1 teaspoon mixed pickling spice
¼ cup garlic-flavored wine vinegar
1 teaspoon salt
¼ teaspoon pepper
1 medium-sized head cabbage, finely shredded
3 tablespoons butter or margarine

Fill a pan with enough water to cover cabbage. Add spice, vinegar, salt, and pepper; bring to a boil. Add cabbage and cook just until tender (about 3 to 5 minutes). Drain and add butter. Makes 8 servings.

## Savory Sauerkraut

1 medium-sized onion, chopped
3 tablespoons bacon drippings
1 can (1 lb.) sauerkraut, drained
1 can (1 lb.) tomatoes
½ teaspoon *each* caraway seed and sugar

Fry chopped onion in bacon fat until limp. Add sauerkraut, tomatoes, caraway seed, and sugar; mix thoroughly.

Turn into a 1 1/2-quart casserole and bake, uncovered, in a 350° oven for 30 to 40 minutes to blend flavors. Makes 6 servings.

## Creamy Cabbage with Peas

1 package (10 oz.) frozen peas, or about 2 cups shelled
  fresh peas
  Boiling salted water
2 tablespoons butter or margarine
¼ teaspoon basil leaves
  Salt and pepper to taste
3 tablespoons water
½ teaspoon salt
6 cups finely shredded cabbage (about 1¼ lbs.)
⅛ teaspoon pepper
½ cup whipping cream

Cook peas in boiling salted water until just tender. Drain; add 1 tablespoon of the butter, basil, and salt and pepper to taste. Cover the pan and keep warm while you cook the cabbage.

Heat the remaining 1 tablespoon butter with the 3 tablespoons water and 1/2 teaspoon salt over highest heat in a 10-inch frying pan with a tight cover. Add the cabbage, cover, and cook 1 minute. Remove cover and add the 1/8 teaspoon pepper and whipping cream. Cook, uncovered, stirring over high heat until the cabbage is tender-crisp and cream is slightly reduced (about 2 minutes). Spoon the cabbage into a warm serving dish, making an indentation in the center. Mound the peas in center of cabbage. Makes 6 to 8 servings.

## Crunchy Cabbage Salad

2 cups finely sliced cabbage
2 cups cooked or canned kidney beans, drained
½ cup chopped salted peanuts
½ cup diced celery
¼ cup diced green pepper
1 tablespoon minced onion
1 cup bottled cole slaw dressing
  Salt

Combine cabbage, beans, peanuts, celery, green pepper, and onion in a salad bowl. Pour dressing over vegetables. Toss salad; add salt to taste and serve. Makes 4 to 6 servings.

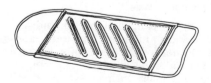

### Cooking and Serving Red Cabbage

When you cook red cabbage, something acid such as vinegar or wine needs to be added to the cooking water to keep the cabbage from turning purple. Red cabbage also requires longer cooking than white cabbage.

### Red Cabbage, German Style

Shred 1 pound head of red cabbage and put in a saucepan with 1/4 cup butter and 1 medium-sized onion, chopped. Cook, covered, for about 10 minutes, until moisture is absorbed. Add 1 tablespoon red wine vinegar and 1 cup dry red wine. Cook until cabbage is tender and liquid reduced (about 20 minutes).

### Red Cabbage, Valencia

Dice 1/2 pound salt pork and cook lightly. Add a 1 1/2-pound head red cabbage, shredded; season with salt, pepper, and dash nutmeg. Add 1/2 cup regular strength chicken broth, 2 tablespoons kirsch or applejack, and 3 apples, peeled and sliced. Cover and cook slowly for 30 minutes; correct seasoning if necessary. This dish can also be garnished with sausages, if desired.

### Red Cabbage, Dutch Style

Shred a 1 1/2-pound head of red cabbage and parboil for 5 minutes; drain. Put 1 medium-sized onion, finely chopped, in a heavy saucepan with 2 tablespoons butter and sauté lightly; add 3 peeled sliced apples, the cabbage, salt and pepper to taste, and, if desired, 1 teaspoon sugar. Simmer 30 minutes. Add 1/2 cup dry red wine, cook quickly, and stir in 1 tablespoon butter before serving.

---

## Red Cabbage in Port

1 small head (about 1 lb.) red cabbage, finely shredded
1 cup Ruby Port
1½ teaspoons vinegar
1 medium-sized apple, peeled, cored, and thinly sliced
Salt

In a large pan, combine cabbage, wine, and vinegar; add apple and bring to a boil. Cover and simmer gently for 1 1/2 hours, stirring occasionally; add a little more wine if cabbage cooks dry. Salt to taste and turn into a serving dish. Makes 2 to 3 servings.

## Bacon-Curry Slaw

6 to 8 cups finely shredded red or green cabbage
⅓ cup mayonnaise
2 tablespoons cider vinegar
1 teaspoon salt
¼ teaspoon *each* pepper and curry powder
1 tablespoon sugar
4 slices bacon, cooked crisp

Place cabbage in a bowl. Stir together the mayonnaise, vinegar, salt, pepper, curry powder, and sugar until smoothly blended. Pour as much dressing as you like over cabbage and mix to coat thoroughly. Crumble bacon over slaw and mix lightly. Serve immediately or cover and refrigerate. Makes 4 to 6 servings.

### Cooking and Serving Celery Cabbage

Celery cabbage is called napa, nappa, or Chinese cabbage in many supermarkets. It can be cooked and served in many of the same ways as green cabbage. You can also serve it raw in salads as you would lettuce. When briefly cooked, it wilts rapidly and develops delicate sweetness.

## Celery Cabbage with Chive Cream Cheese Sauce

2 tablespoons butter
1 medium-sized head (about 2 lbs.) celery cabbage cut in 1-inch wide pieces (about 12 cups)
1 small package (3 oz.) cream cheese with chives
Salt and pepper to taste

In a wide frying pan, melt butter and add cabbage. Cover and cook over high heat just until cabbage begins to wilt. Stir as needed to cook evenly. Remove pan from heat and push a small area clear in pan; place cheese in it and quickly mash with a spoon until soft. Tilt pan to drain juices into the cheese, blending smoothly; then mix with cabbage. Season with salt and pepper and serve at once. Makes 4 to 5 servings.

## Celery Cabbage with Sour Cream and Cheese Dressing

8 cups (about ⅔ of a 2-lb. head) thinly sliced, well-crisped, celery cabbage
½ cup sour cream (or sour cream chive dressing)
1½ tablespoons dry onion soup mix
1 ounce (about 1½ tablespoons) blue cheese
½ cup mayonnaise
1 tablespoon minced parsley
2 to 3 tablespoons milk or half-and-half (light cream)
2 to 3 tablespoons finely chopped green onions, including part of tops

Chill celery cabbage while you prepare the dressing. Beat together in a bowl the sour cream, dry onion soup mix, blue cheese, mayonnaise, and minced parsley. Thin dressing to a good mixing consistency with the milk or half-and-half.

Pour the cheese dressing over the cabbage in a salad bowl and scatter the chopped green onions over the dressing. Mix well and serve on individual salad plates at the table, if desired. Makes 6 to 8 servings.

# CARROTS

## Selection and Storage

**Selection:** Choose firm, clean, well-shaped, smooth carrots with bright orange-gold color and fresh green tops. Young, sweet, immature carrots have long rootlets.

**Buying:** Allow 1 pound or 1 medium-sized bunch for each 3 to 4 servings.

**Storage:** Remove tops from bunched carrots, put in plastic bags; close top tightly. Or leave in plastic bag purchased in. Store in refrigerator. Will hold 1 to 2 weeks.

## Cutting and Cooking

Young, unblemished carrots can be simply scrubbed with a vegetable brush and rinsed. If necessary, scrape carrots or remove a thin outer layer with vegetable peeler. Carrots can be cooked whole, sliced lengthwise in halves or quarters, cut into sticks, cut in chunks, cut in thin or thick slices, chopped, or shredded.

**To cook carrots,** you can use a small amount of water (about 1/2 cup) with heavy pans that have tight-fitting lids, or drop the cut carrots into 1/2 to 1 inch boiling salted water, or use a steamer. Cover and cook gently or steam until just tender. Cooking times depend on maturity of carrots and the cut: whole baby carrots take 8 to 10 minutes; larger, more mature whole carrots take 10 to 20 minutes; cut, sliced, or shredded carrots cook in 4 to 10 minutes. Drain immediately unless water is all absorbed.

## Seasoning and Serving

Young, sweet, cooked carrots need only be seasoned with salt, pepper, and a little butter. In addition, you might add a pinch of sugar; a dash of dill weed, curry powder, ground nutmeg, ground ginger; a few crumbled leaves of basil, mint, thyme, or marjoram; a few drops lemon juice; or chopped parsley.

**If you cook the carrots ahead,** cook them until just tender, drain, cool in cold water, drain again.

The following seasoning suggestions are based on about 1 bunch medium-sized carrots, or 1 pound, cut and cooked as directed above.

### Carrots Julienne

Cut carrots into matchstick-size pieces and cook as directed above; drain. Add to the pan 1/4 cup whipping cream and 3 tablespoons minced chives. Heat quickly to boiling, and serve.

### Mashed Carrots

Cut and cook carrots any of the ways suggested under *cutting and cooking* above. Cook until fully tender; drain. Mash as you would for potatoes, adding butter, salt, and pepper to taste.

### Browned Butter Carrots

Cut carrots lengthwise in halves, cook as directed under *cutting and cooking* above; drain in a colander. In the pan heat about 4 tablespoons butter until it starts to brown; add carrots and salt and pepper to taste. Sauté until carrots are hot, and turn into serving dish. Rinse the pan with 2 tablespoons Sherry and distribute over vegetables.

### Carrots in Cream

Cut carrots into slices or use tiny, young carrots whole; cook as directed under *cutting and cooking* above; drain well. Season to taste with salt, pepper, and a pinch sugar. Add to the pan 1/2 cup whipping cream, bring to boiling, and boil until thickened slightly (about 3 minutes). Serve.

### Carrots Cointreau

Cut carrots in thin, slanting slices; cook as directed under *cutting and cooking* above; drain. Add to the pan 2 tablespoons *each* butter and Cointreau (or other orange-flavored liqueur); stir until butter is melted and carrots glazed.

### Brandied Carrots

Use small whole carrots or larger carrots, cut in halves or quarters; cook as directed under *cutting and cooking*; drain. In the pan melt 2 tablespoons butter. Add 1 1/2 tablespoons brandy, and when warm, set aflame. Shake pan back and forth until carrots are lightly browned all over. Serve.

### Glazed Carrots

Use small, young carrots whole or cut larger carrots in 1/2-inch-thick slanting slices; cook as directed on page 31; drain. Melt in the pan 3 tablespoons butter, stir in 3 tablespoons granulated or brown sugar. Tip pan and turn carrots over high heat until glazed all over. Sprinkle with parsley when served.

---

### Butter-Steamed Carrots

Melt 2 tablespoons butter in a wide electric frying pan set at high heat (or use a wide frying pan over direct high heat). Add 3 cups thinly sliced, peeled carrots (10 or 12 slender carrots) and 3 tablespoons water. Cover and cook, stirring occasionally, for 5 minutes. Season with salt and pepper to taste. Makes 4 to 5 servings.

### Cream-Glazed Butter-Steamed Anise Carrots

Prepare the carrots as directed for Butter-Steamed Carrots above, adding 1/4 teaspoon anise seed along with the carrots. After 5 minutes, add 3 tablespoons whipping cream and cook, uncovered, stirring, until liquid is almost all evaporated. Makes 4 to 5 servings.

## Saucy Carrots

4 tablespoons tomato sauce (half an 8-oz. can)
1 yellow onion, sliced crosswise and then divided into rings
2 cups small carrots cut into about 1-inch lengths and cooked until barely tender
Pinch basil leaves

Place tomato sauce in pan and add onion rings. Simmer just until the onions are partially cooked but not too crunchy. Drain carrots, add to onions and sauce, and sprinkle with basil. Add salt, if desired. Makes 3 to 4 servings.

## Ginger-Minted Appetizer

3 packages (10 oz. *each*) frozen baby carrots
1 cup orange juice
1 teaspoon grated fresh ginger root
Dash *each* salt and pepper
1 tablespoon chopped fresh mint

Combine the carrots, orange juice, ginger root, salt, and pepper in a pan. Cover and bring to boil; simmer until carrots are just tender (about 3 minutes). Chill carrots, covered in liquid. Drain and spoon into serving bowl; garnish carrots with fresh mint. Serve with picks. Makes about 6 cups appetizers.

## Marinated Carrots with Artichokes

1 bunch carrots (about 5 medium-sized carrots)
⅓ cup olive oil (may be part salad oil)
⅔ cup water
2 tablespoons lemon juice
½ teaspoon grated lemon peel
2 teaspoons salt
1 teaspoon sugar
⅛ teaspoon pepper
1 package (8 oz.) frozen artichoke hearts
Lettuce for garnish (optional)

Cut carrots into 1/2-inch-thick slanting slices. Heat olive oil in a large pan, add carrots, cover, and cook until partially tender (about 8 minutes). Add the water, lemon juice and peel, salt, sugar, and pepper. Bring to boiling, add the artichoke hearts, and cook until they are just tender (about 4 minutes). Lift out vegetables with a slotted spoon and arrange in a bowl. Cool vegetables and the cooking liquid separately, then pour liquid over vegetables and refrigerate for several hours.

To serve, lift vegetables out of the marinade and arrange on serving plate; spoon some of the marinade over top. Garnish with lettuce leaves, if you wish. Makes 4 to 6 servings.

## Cinnamon Carrot Sticks

12 large carrots (2 to 2½ lbs.), peeled and cut into sticks about ½ by 3 inches
2½ cups water
½ cup cider vinegar
¾ cup sugar
1 stick whole cinnamon

Place carrots in a large pan with water and boil, covered, until just tender when pierced (7 to 10 minutes); drain, saving liquid. Turn carrots into a bowl. To carrot liquid add the cider vinegar, sugar, and cinnamon; bring to a boil and pour over carrots. Cool, cover, and chill overnight or as long as 2 to 3 weeks. Serve as an appetizer or relish. Makes about 7 cups.

## Oriental-Style Carrots

3 to 4 cups sliced carrots (cut slanting slices about ½ inch thick)
Boiling salted water
About 1 tablespoon salad oil
1 or 2 teaspoons finely chopped fresh ginger root (optional)
1 or 2 cloves garlic, minced
About 1 tablespoon water
1 teaspoon sugar
½ teaspoon salt
Chopped parsley or toasted sesame seed for garnish

Drop carrots into boiling water and pre-cook until just tender (about 4 to 5 minutes). Drain, cool quickly, drain again, and set aside. Measure and prepare all remaining ingredients, and have within reach of your range.

Heat a 10-inch or larger frying pan or wok over high heat. Put in oil. As soon as oil is hot enough to ripple when pan is tipped, put in ginger and garlic; quickly stir with a spatula until it starts to brown, about 30 seconds. Put in carrots, water, sugar, and salt. Keep turning with a spatula until heated through (about 1 to 2 minutes). Turn into a warm serving dish; garnish top with parsley or sesame seed. Serves 4 to 5.

## Gingered Carrots

Prepare Oriental-Style Carrots as directed on page 32, omitting fresh ginger. Stir 1 tablespoon minced candied ginger into the carrots.

## Carrots with Peas

Quickly defrost 1/2 to 1 package (10 oz. size) frozen peas by holding in a colander under hot tap water; drain. Follow recipe above for Oriental-Style Carrots, reducing the amount of carrots used to 2 to 3 cups; stir-fry the two vegetables together.

## Carrot Onion Ring Salad

10 large carrots (about 1¾ lbs.)
   Boiling salted water
1 medium-sized white onion, sliced
½ cup salad oil
¼ cup white wine vinegar
2 tablespoons sugar
1 teaspoon celery seed
¼ teaspoon dry mustard
½ teaspoon salt
⅛ teaspoon paprika

Peel carrots and cut slanting slices about 1/4 inch thick. Cook, covered, in 1 inch boiling water until just tender (about 5 minutes); drain. Plunge into cold water, drain again, and turn into a serving bowl. Separate onion slices into rings; add to carrots. In a small jar, combine the oil, vinegar, sugar, celery seed, mustard, salt, and paprika. Shake to blend; pour over carrots. Cover and chill overnight, stirring several times. Makes 6 to 8 servings.

## Golden Carrot Soup

2 cups peeled, sliced carrots (about 4 carrots)
1 cup boiling water
½ teaspoon salt
2 tablespoons chopped onion
1½ teaspoons chopped fresh mint (or ¾ teaspoon dry mint)
2 tablespoons melted butter or margarine
2 tablespoons all-purpose flour
2 cups regular strength chicken broth or milk
¼ teaspoon ground nutmeg
   Salt to taste
¾ cup orange juice
½ cup whipping cream
¼ teaspoon grated orange peel
   Dash ground nutmeg

In a saucepan cook carrots in water and salt until tender. Whirl carrots and cooking liquid smooth in a blender. In another saucepan sauté onion and mint in butter until soft. Mix in flour and stir until bubbly; gradually add chicken broth or milk, the carrot mixture, and the 1/4 teaspoon nutmeg. Simmer over medium-low heat, stirring, for 3 to 4 minutes. Add salt to taste and stir in orange juice.

   Pour into heat-proof soup bowls. Lightly whip cream, add grated orange peel, and a dash of nutmeg. Spoon even portions of cream onto each bowl of soup; set under broiler until top is nicely browned. Makes 4 to 6 servings.

## Carrots in Wine Sauce

8 medium-sized carrots, peeled
3 tablespoons butter or margarine
1 bunch green onions, including part of tops, thinly sliced
¼ teaspoon salt
1 tablespoon water
4 teaspoons all-purpose flour
⅔ cup half-and-half or milk
3 tablespoons dry white wine, Sherry, or Madeira
   Finely chopped parsley

Slice carrots about 1/8 inch thick. In a frying pan that has a tight cover, heat the butter. Add the carrots and onions and sauté about 3 minutes. Add the salt and water; cover and simmer until carrots are almost tender (about 7 minutes). Sprinkle with flour and cook, stirring, until bubbly. Remove from heat and gradually stir in the cream and wine. Then cook, stirring, until thickened. Serve immediately. (Carrots can be refrigerated at this point.) Just before serving, reheat over low heat, stirring several times just until heated through. Spoon into a serving dish and sprinkle with parsley. Makes 4 to 6 servings.

## Carrots in Orange Sauce

Follow directions for Carrots in Wine Sauce above using 3 tablespoons orange juice with 1 teaspoon grated orange peel in place of the wine.

## Carrot Soufflé

2 cups packed shredded carrots
1 tablespoon melted butter or margarine
3 tablespoons water
1 teaspoon salt
1 tablespoon minced chives
4 tablespoons all-purpose flour
1 cup skim milk
4 eggs, separated

In a heavy saucepan, combine the carrots, butter, water, salt, and chives. Cover and cook over medium heat, stirring occasionally, until the carrots are tender, (about 5 minutes). Sprinkle with flour and stir until mixed. Slowly add the milk, stirring, and cook until thickened. Remove from heat. Beat in 4 egg yolks. Whip the 4 egg whites until they hold soft peaks; fold into carrot mixture. Pour into a deep buttered casserole (1 1/2 to 2-qt. size). Bake in a 375° oven 35 minutes or until puffed and lightly browned. Serve immediately. Makes 5 to 6 servings.

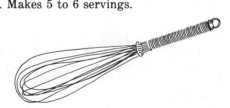

# CAULIFLOWER

## Selection and Storage

**Selection:** Choose white or creamy white compact heads of tight flowerets that are free of discolored spots. Yellow tinge or spreading flowerets indicates over-maturity. If leaves are attached, they should be green and crisp.

**Buying:** Allow 1 medium-sized head (1 1/4 lb.) for each 4 to 6 servings.

**Storage:** Place, unwashed, in a plastic bag and seal, or leave in plastic wrap purchased in. Store in the refrigerator. Use as soon as possible.

## Cutting and Cooking

Wash thoroughly. Remove outer stalk leaves—if they are fresh looking, they'll be good cooked with the cauliflower or as a separate vegetable. The head can be left whole or separated into flowerets, discarding the core. Also flowerets may be sliced lengthwise through flower and stem, or may be chopped. Be very careful not to overcook cauliflower, as it develops strong flavor and odor.

**To cook cauliflower,** place in 1-inch boiling salted water (use less water with heavy pans). Cover and cook until just tender-crisp; whole head takes 15 to 30 minutes, depending on size; flowerets take 5 to 9 minutes; sliced or chopped take 3 to 5 minutes; drain.

**When cooking cauliflower in advance** plunge immediately into ice water after cooking to stop the cooking, then drain again. Chill until needed.

## Seasoning and Serving

Cauliflower can be delicious served raw or cooked. Raw, it can be served whole with a small sharp knife for cutting it apart, or cut into individual flowerets. Offer it as an appetizer for dipping, along with other raw vegetables or as a vegetable relish.

Lightly cooked, hot cauliflower can be served with melted butter or with butter to which minced parsley, chopped chives, or dill weed has been added. It is nicely complemented with Mornay, Hollandaise, Béchamel, or Pesto Sauce (see pages 6 - 7). Each one of the following suggestions is based on a medium-sized cauliflower making enough for 4 to 6 servings.

### Cauliflower Casserole

Cut cauliflower into flowerets and cook as directed above; drain well. Put into a shallow casserole and cover with about 1 1/2 cups Mornay Sauce (see page 6 ). Sprinkle with about 3 tablespoons freshly grated Parmesan or Romano cheese. Place about 5 inches below heat in broiler, until lightly browned (4 to 5 minutes).

### Cauliflower with Chive Hollandaise

Cook the cauliflower whole or cut apart in flowerets, following directions above; drain. Arrange on a serving plate or in a bowl. In a small pan combine 1 can (6 oz.) Hollandaise Sauce, 1/4 cup sour cream, and 1 tablespoon lemon juice. Stir over medium heat until heated through. Pour over cauliflower. Sprinkle with about 1/2 tablespoon minced chives.

### Cauliflower with Cheese and Green Onion

Cut cauliflower into flowerets and cook as directed above; drain. Combine 1 jar (8 oz.) pasteurized process cheese spread with 1/4 teaspoon garlic salt and dash pepper. Add to the hot cauliflower with 2 or 3 green onions, including part of the tops, sliced. Mix lightly until blended.

---

### Butter-Steamed Cauliflower

Melt 3 tablespoons butter in an electric frying pan set at highest heat or use a wide frying pan over direct high heat. Add 4 cups thinly sliced cauliflower and 6 tablespoons water. Cover and cook, stirring occasionally, for 5 minutes. Season with salt and pepper to taste. Makes 4 to 6 servings.

### Cauliflower with Egg Sauce

Prepare cauliflower as directed for butter-steaming above. When cooked, pour into serving bowl and blend with 2 finely chopped hard-cooked eggs and 1/2 cup warm Hollandaise Sauce (use canned Hollandaise or one of the recipes, page 6). Makes 6 servings.

## Cauliflower à la Polonaise

1 medium-sized head cauliflower
　Boiling salted water
½ cup (¼ lb.) butter or margarine
⅓ cup fine dry bread crumbs
1 hard-cooked egg, chopped
　Parsley sprigs

Cut out thick core from cauliflower; break into flowerets and wash. Cook as directed under *cutting and cooking*, page 34; drain and arrange in a serving dish.

Heat butter in a saucepan until bubbly; pour over cauliflower, leaving about 3 tablespoons remaining in saucepan. Stir crumbs into remaining butter until combined; add to cauliflower with the chopped egg. Using 2 forks, mix lightly until well combined. Garnish with parsley; serve at once. Makes 4 servings.

## Oriental-Style Cauliflower

　About 4 cups sliced cauliflower
　Boiling salted water
1 tablespoon salad oil
1 or 2 cloves garlic, minced or mashed
1 tablespoon water
1 teaspoon *each* sugar and dill weed (optional)
　About ½ teaspoon salt
　Chopped parsley or sliced green onion for garnish

Divide cauliflower into flowerets, then slice lengthwise through flowers to make slices about 1/4 inch thick. Precook in the boiling water just until barely tender when pierced with a fork (2 to 3 minutes). Drain, cool quickly with cold water, and set aside.

Before starting to cook, prepare all remaining ingredients and have them within reach of your range. Heat a 10-inch or larger frying pan or a wok over high heat, then put in oil. As soon as oil is hot enough to ripple when the pan is tipped, add garlic; quickly stir with a wide spatula until browning starts (about 30 seconds). Put in cauliflower, water, sugar, dill weed (if used), and salt. Keep turning with a spatula until heated (about 1 to 2 minutes). Turn out into a warm serving dish and sprinkle with parsley or green onion, if desired. Makes 4 to 5 servings.

## Cauliflower Cheese-Dilly

1 medium-sized head of cauliflower (about 1¼ lbs.)
4 tablespoons (⅛ lb.) butter or margarine, melted
4 tablespoons grated Parmesan cheese
1 teaspoon dill weed
1 clove garlic, minced or mashed (or ⅛ teaspoon garlic
　powder)
　Salt and pepper

Cut cauliflower into flowerets and cook following directions under *cutting and cooking*, page 34; drain, and turn into a 1-quart baking dish. To the butter add 2 tablespoons of the Parmesan cheese, the dill weed, and garlic; mix with the hot cauliflower. Season to taste with salt and pepper. Sprinkle with the remaining 2 tablespoons grated Parmesan. Place under broiler about 6 inches from heat for about 10 minutes or until the top is golden brown. Makes 4 to 6 servings.

## Cream of Cauliflower Soup

2 tablespoons butter or margarine
1 large onion, chopped
2 cans (14 oz. *each*) regular strength chicken broth
2 medium-sized carrots
1 medium-sized cauliflower
1 cup half-and-half (light cream)
⅛ teaspoon ground nutmeg
　Salt and pepper
1 tablespoon dry Sherry (optional)
1 tablespoon chopped parsley

In a 3 or 4-quart saucepan melt butter over medium heat; add onion and sauté until limp (about 5 minutes). Pour in broth and bring to a boil.

Meanwhile, cut carrots into 1/4-inch slices and cut cauliflower into small flowerets. Add vegetables to boiling broth, reduce heat, cover, and simmer until vegetables are tender when pierced (about 7 minutes). Pour a small amount at a time into a blender container and whirl until smooth. Turn all the purée into a pan, add cream, nutmeg, salt and pepper to taste, and Sherry, if desired. Heat to simmering. Serve garnished with chopped parsley. Makes 4 to 6 servings.

## Chilled Cauliflower with Guacamole

1 medium-sized cauliflower
2 tablespoons salad oil
　Salt and pepper to taste
1 avocado mashed (about 1 cup)
4 tablespoons *each* lemon juice, minced canned California
　green chiles, and chopped green onion
¾ teaspoon salt
¼ cup sour cream
　Dash liquid hot-pepper seasoning
　Crisp romaine lettuce
　Radish roses

Cook cauliflower whole following directions under *cutting and cooking*, page 34; drain. Sprinkle with salad oil and salt and pepper to taste; chill.

Combine the avocado, lemon juice, green chiles, green onion, 3/4 teaspoon salt, sour cream, and hot-pepper seasoning in a blender. Whirl until smooth. Just before serving, put cauliflower on a large platter and surround with romaine lettuce, cover the cauliflower with avocado mixture, and garnish with radish roses. Makes 6 to 8 servings.

## Cauliflower, Pepper and Peas

1 teaspoon olive oil
1 green or red bell pepper, stemmed, seeded, and cut into
　12 pieces
2 cups cauliflower cut in small flowerets
1 package (10 oz.) frozen peas
¾ cup water
⅛ teaspoon ground cumin seed
¼ teaspoon salt

Heat oil in a wide frying pan. Stir in pepper, cauliflower, peas, water, cumin, and salt. Cover and cook over moderately-high heat for 8 to 10 minutes or until cauliflower is just tender. Makes 4 servings.

## Carrot and Cauliflower Medley

About 8 large carrots, peeled
1 large head cauliflower
Boiling salted water
2 tablespoons *each* butter or margarine and all-purpose flour
¼ teaspoon prepared Dijon-style mustard
1 cup regular strength chicken broth
½ cup whipping cream
1¼ cups shredded Swiss cheese
2 green onions, including part of tops, sliced

Cut carrots into 1/4-inch-thick slanting slices (you should have 4 cups). Break cauliflower into flowerets. Cook vegetables in boiling salted water until just tender (about 5 minutes). Plunge in cold water; drain.

In another pan, melt butter over medium heat; stir in flour and mustard and cook until bubbly. Remove from heat; gradually stir in chicken broth and cream. Cook, stirring, until thickened. Gradually add 1 cup of the cheese, stirring, until melted.

Combine vegetables and sauce in a 2-quart casserole. Sprinkle with remaining cheese. If done ahead, cover and refrigerate.

Before serving, bake, uncovered, in a 350° oven until heated through (15 minutes if warm or 35 minutes if chilled). Garnish with onions. Makes 8 servings.

## Green and White Salad Platter

1 large cauliflower
Boiling salted water
2 packages (8 or 9 oz. *each*) frozen artichoke hearts
½ cup white wine vinegar
¾ cup salad oil
½ teaspoon *each* dry mustard, garlic salt, and basil leaves, crumbled
⅛ teaspoon pepper
1 tablespoon instant minced onion
Crisp salad greens
Cherry tomato halves

Cut cauliflower into flowerets and cook following directions under *cutting and cooking*, page 34; drain. Immerse in cold water, drain again; put in a bowl. Cook artichokes according to package directions; immerse in cold water, drain, and place in another bowl.

Meanwhile, in a small bowl or jar, combine the vinegar, oil, mustard, garlic salt, basil, pepper, and onion. Stir or shake to blend. Pour half of the dressing over each vegetable. Mix gently, cover, and chill overnight, stirring several times.

Before serving, line a serving platter with lettuce. Lift vegetables from marinade with a slotted spoon and arrange on lettuce. Garnish with tomatoes. Serve with remaining marinade. Makes about 8 servings.

 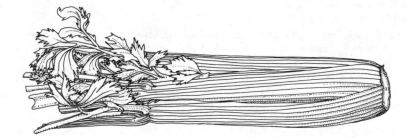

## Selection and Storage

**Selection:** Choose celery with rigid, crisp stalks and fresh looking leaves. Avoid rubbery wilted looking branches.

**Buying:** 1 medium bunch will serve 4 to 6; 1 heart will serve 2.

**Storage:** Rinse bunch. Place in plastic bag, close tightly to keep in moisture, and refrigerate. Will keep 1 to 2 weeks.

## Cutting and Cooking

When ready to use separate stalks (or branches) from bunch. Scrub stalks with a vegetable brush, trim off root, leaves, and any blemishes. Or, for celery hearts, split whole heart in half lengthwise, trim off all but tender leaves, and wash.

The stalks can be cut in thin or thick slices, in long slanting pieces, cut in long strips, or diced.

**To cook pieces,** place in a saucepan of 1-inch boiling salted water (or use less water with heavy pans). Cook, covered, until tender-crisp (5 to 10 minutes), depending on size of pieces; drain.

**To cook celery hearts,** add them to 1-inch or more boiling salted water. Cover and cook 10 to 20 minutes or until just tender; drain.

## Seasoning and Serving

Celery, most popular served raw, can be a distinguished addition to the menu when cooked. Celery hearts, split lengthwise in halves, make an especially handsome presentation as a cooked vegetable. Either the hearts or cut pieces of celery are delicious when cooked in broth or consommé instead of water. Basil, tarragon, and thyme leaves can be added to the cooking water or broth when cooking celery. Or the cooked celery can be served with Mornay, Béarnaise or Hollandaise Sauce.

---

### Butter-Steamed Celery

Melt 2 tablespoons butter in a wide electric frying pan set at highest heat or use a wide frying pan over direct high heat. Add 4 cups thinly sliced celery and 2 tablespoons water. Cover and cook, stirring occasionally, for 3 minutes. Salt to taste. Makes 4 to 6 servings.

### Butter-Steamed Parmesan Celery

Prepare celery as directed for butter-steaming above. Just before serving, sprinkle with 4 tablespoons shredded Parmesan cheese. Serves 4 to 6.

### Lemon-Cheese Celery

Prepare celery as directed for butter-steaming above. Remove cover after cooking and stir in salt to taste, 1 teaspoon lemon juice, 1/4 cup shredded sharp Cheddar cheese, and a dash of cayenne. Continue heating just long enough to melt the cheese; serve immediately. Makes 4 to 6 servings.

## *Celery Appetizers*

Here are some simple ways to use raw celery as appetizers and snacks.

**Celery Fans:** Cut large stalks of celery in sections about 2 inches long. Then cut each section into thin strips, cutting lengthwise to within 1/2 inch of end. Cover with ice water and chill for several hours, or until celery opens into little fans. Drain and serve.

**Shrimp-Stuffed Celery:** Combine 1 small (3 oz.) package cream cheese (softened) with about 1 cup finely chopped or ground cooked shrimp, and enough mayonnaise to make a good spreading consistency. Spread on small pieces of celery.

**Cheese-Stuffed Celery:** Blend 1 jar (5 oz.) pasteurized processed American cheese spread with about 1 tablespoon sweet pickle relish. Or use prepared cream cheese with olives and pimientos or another flavored cream cheese to stuff small, tender pieces of celery.

**Celery Sandwich Rolls:** Trim crusts from thinly sliced fresh, white bread. Spread each with softened butter. Roll a small, tender stalk of celery in each slice. Pack rolls, seam sides down, in a box or dish, with waxed paper between layers. Refrigerate until well chilled.

## Celery Hearts with Spiced Tomato Sauce

    6 whole celery hearts, trimmed to same length, all leaves
        cut off
      Boiling water
    2 chicken bouillon cubes
    1 teaspoon salt
    1 garlic clove, cut
    ½ small clove garlic, minced
    1 cup catsup
    2½ tablespoons salad oil
    1 tablespoon wine vinegar
    1½ teaspoons celery seed
    ¼ teaspoon salt

Tie a string around the top of each heart to hold it together. In a wide shallow pan, heat enough boiling water to almost cover celery hearts. Add bouillon cubes and the 1 teaspoon salt. Add celery and cook, covered, for 10 to 12 minutes or until just barely tender. Chill in stock—overnight, if desired.

Rub the cut clove of garlic around a small bowl. In the bowl, mix the minced garlic, catsup, salad oil, wine vinegar, celery seed, and the 1/4 teaspoon salt.

Remove strings from celery, arrange on a serving dish, and drizzle some of the sauce over all. Offer extra sauce to ladle over each heart. Makes 6 servings.

## Braised Celery Hearts

    3 celery hearts
    4 tablespoons butter or margarine
    ¼ teaspoon thyme leaves
      Dash pepper
    1 can (10 oz.) condensed consommé
      Salt to taste

Split celery hearts in half lengthwise and trim off all but tender leaves. In a 10-inch frying pan (one with a tight fitting lid) melt butter over medium heat. Place celery in pan with cut side down and sauté until browned. Sprinkle with thyme and dash pepper. Add consommé, cover, and simmer until tender (10 to 20 minutes). Add salt, if needed; drain. Makes 6 servings.

## Peas and Celery Medley

    2 tablespoons butter or margarine
    2 cups sliced celery
    3 tablespoons water
    2 packages (10 oz. *each*) frozen peas
    1 can (10½ oz.) condensed cream of mushroom soup
    1 can (3 or 4 oz.) sliced mushrooms
    1 can (5 oz.) water chestnuts, drained and sliced
    1 tablespoon butter, melted
    ¾ cup fresh bread crumbs

Melt the 2 tablespoons butter in a pan. Add celery and water, cover tightly and cook, stirring occasionally until tender-crisp (about 4 minutes). Add peas and simmer 5 minutes more. Heat soup with the mushrooms and their liquid and water chestnuts. Layer vegetables alternately with creamed mixture into a buttered 1 1/2-quart casserole. Mix the 1 tablespoon butter with bread crumbs; sprinkle over top. Bake in a 350° oven for 20 minutes or until bubbly. Makes 4 servings.

## Celery Avocado Green Salad

    5 large stalks celery, cut in slanting slices
    ½ cup regular strength chicken broth
      Victor Dressing (recipe follows)
    1 head butter lettuce
    ½ bunch watercress or ¼ head chicory (curly endive)
    1 avocado, peeled and cut in chunks
      Minced parsley and pimiento strips for garnish

In a saucepan, cook celery in the chicken broth until almost tender (about 4 minutes); drain. While celery is still warm, pour over the Victor Dressing. Cover and refrigerate for about 6 hours or overnight. To serve, tear lettuce and watercress into a shallow bowl. Lift celery out of dressing with a slotted spoon and arrange with the avocado on top of greens. Garnish with parsley and pimiento. Drizzle remaining dressing over salad. Bring to table and mix. Makes about 6 servings.

**Victor Dressing.** Combine in a bowl or jar 1/2 cup salad oil, 2 tablespoons *each* lemon juice and minced dill pickle, 1 tablespoon *each* minced onion and minced capers, 1 teaspoon salt, and 1/4 teaspoon dry mustard.

## Celery with Tarragon

1 bunch celery, stalks cut into 1-inch-size pieces
1 teaspoon tarragon leaves
¼ cup butter or margarine
1 cup water

Place celery in a saucepan with tarragon, butter, and water. Cover tightly and simmer until barely tender (about 5 minutes). Uncover and cook quickly over high heat until all liquid has evaporated. Add salt, if desired, and serve. Makes 4 to 6 servings.

## Celery with Water Chestnuts

1 bunch celery
1 can (14 oz.) regular strength chicken broth
½ teaspoon basil leaves
½ teaspoon salt
¼ teaspoon pepper
2 tablespoons *each* cornstarch and cold water
1 can (12 oz.) water chestnuts
⅓ cup sliced almonds
½ cup fine dry bread crumbs
3 tablespoons melted butter or margarine

Trim celery and slice into 1-inch pieces. Put into a large saucepan, add chicken broth, basil, salt and pepper. Cover and cook until celery is tender-crisp (about 10 minutes).

Blend cornstarch with water and gradually stir into hot stock; cook, stirring, just until thickened. Drain and slice water chestnuts; add to sauce along with almonds. Turn into a 2-quart casserole. Mix bread crumbs with melted butter and sprinkle over top of vegetable. Bake, uncovered, in a 350° oven for 30 minutes. Makes about 6 servings.

# CELERY ROOT

## Selection and Storage

**Selection:** Choose celery root (celeriac) with firm, clean roots.

**Buying:** 1 pound serves about 2 to 3.

**Storage:** Keep unwashed celery root tightly wrapped in the refrigerator to retain moisture. Will keep 1 week or more.

## Cutting and Cooking

The cut surface of raw celery root darkens quickly when exposed to air. To retain its milky white color, don't cut it until just before cooking, or cook the whole root, unpeeled, then peel and cut it after cooking.

For salads, celery root is often cooked briefly to slightly soften it and keep it from darkening. When you serve celery root raw, rub the cut surfaces with full-strength or slightly diluted lemon juice or vinegar.

**To cook whole celery root,** scrub the root thoroughly with a vegetable brush to remove all dirt. Cut off leaves and rootlets. Put into boiling salted water to cover; cook, covered, until tender (40 to 60 minutes). When cool enough to handle, peel and cut as desired.

**To cook cut-pieces,** cut off top and root and peel away the thick outer skin. Cut into slices, match-stick-size pieces, dice, or shred. Immediately drop into about 1 inch boiling salted water; cover and cook until just tender (5 to 15 minutes), depending on size of pieces; drain.

## Seasoning and Serving

Celery root cooked until tender-crisp can be delicious served with melted butter and a sprinkling of minced parsley. Other seasonings that enhance it are basil leaves, onion, oregano leaves, and tarragon leaves. The following suggestions are based on 2 roots (about 2 lbs.) or the amount for 4 to 6 servings.

### Sautéed Celery Root

Trim and peel roots as directed above; cut each into eighths. Cook in boiling salted water until partially tender (about 10 minutes); drain. Add 1/4 cup (1/8 lb.) butter to the pan. Cover and finish cooking about 10 minutes longer. Add salt and pepper to taste, and sprinkle with minced parsley.

### Celery Root Parmesan

Pare and dice celery root as directed above. Instead of cooking in water, drop the cut pieces into boiling bouillon or beef stock to cover until tender (10 to 15 minutes). Drain and serve at once, sprinkled with freshly grated Parmesan cheese.

### Celery Root in Cream

Pare and dice celery root and cook as directed above until just tender; drain. Add to the pan 1/3 cup whipping cream and 3/4 teaspoon celery seed. Simmer rapidly, stirring, until sauce thickens and lightly coats. Add salt and pepper to taste.

### Butter-Steamed Celery Root

Melt 4 tablespoons butter in an electric frying pan over high heat or in a wide frying pan over highest heat. Add 3 cups coarsely shredded celery root and 2 tablespoons water. Stir and cover. Cook, stirring occasionally, until tender (about 5 minutes). Remove cover, season to taste with salt and pepper, and sauté until all liquid has evaporated. Sprinkle with minced parsley or Parmesan cheese. Makes 4 to 6 servings.

## Beef and Celery Root Soup

3 pounds sliced beef shanks
1 tablespoon salad oil or olive oil
1 onion, chopped
1 carrot, chopped
1 teaspoon salt
3 quarts water
3 tablespoons beef stock base or 9 beef bouillon cubes
2 celery roots (about 2 lbs.), pared and diced in bite-size pieces
1 bunch leeks
  Freshly shredded Parmesan cheese

Using a large Dutch oven or soup kettle, brown meat in oil on both sides. Push meat to the sides of the pan, add onion and carrots, and sauté until vegetables are limp, stirring occasionally. Season with salt and pour in water. Cover and simmer slowly for 2 to 2 1/2 hours or until meat is tender. Remove meat from stock and let cool slightly; discard bones and fat and dice meat. Skim any fat from stock and bring to a boil; add beef stock base.

Add celery root to stock, and simmer 10 minutes. Trim all the coarse fibrous leaves from leeks and halve lengthwise; wash well and slice. Add to broth along with meat and simmer 5 minutes longer. Pass cheese at the table to sprinkle over servings. Makes 6 to 8 servings.

## Celery Root and Avocado

1 egg yolk
1 teaspoon prepared Dijon-style mustard
3 tablespoons lemon juice
¼ teaspoon onion salt
⅛ teaspoon pepper
¼ cup whipping cream
1 celery root (about 1 lb.), pared
⅓ cup white wine vinegar mixed with 2 tablespoons water
1 large fully-ripe avocado
4 large lettuce leaves

Stir together in a bowl the egg yolk, mustard, 1 tablespoon of the lemon juice, onion salt, and pepper until smooth; gradually stir in cream until smooth; cover.

Cut celery root into quarters vertically and then shred into a bowl. Add vinegar-water mixture and toss with hands to distribute over the celery root. Transfer celery root to colander; press down firmly with back of a spoon to remove the vinegar-water. Mix drained celery root with prepared sauce in the bowl until sauce is thoroughly distributed.

Peel, remove seed, and slice avocado; sprinkle remaining 2 tablespoons lemon juice over cut surfaces. Place a lettuce leaf on each of four salad plates; divide avocado slices between the plates and arrange on lettuce. Last, spoon shredded celery root onto avocado and serve at once. Makes 4 servings.

 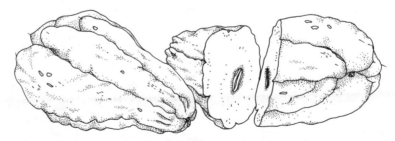

# CHAYOTE

### Selection and Storage

**Selection:** In shape and size, a chayote resembles a papaya, but it is used more like a vegetable. Its color varies from almost white to dark green, and its flesh is firm and rather crisp. Choose one that is firm and free from blemishes.

**Buying:** A medium-sized chayote (about 14 oz.) makes 3 to 4 servings.

**Storage:** Because it is a tropical fruit, chayote keeps best in a cool (not cold) place. If possible, store it at a temperature slightly above 50°; it keeps well stored like this for 2 to 3 months.

### Cutting and Cooking

If the skin is tender, you can cook a chayote without peeling it. Inside is a large flat seed that can be eaten too—it has a nutlike flavor. Scrub the chayote well or peel it. Then dice or slice it (you can slice right through the seed). With the seed lifted out, chayote shells are a good size and shape for stuffing and baking. For this, the shells are usually precooked.

**To precook chayote halves,** place them in a large amount of boiling salted water, cover, and cook until almost tender (30 to 40 minutes); drain.

**To cook sliced or diced chayote,** place in a pan with a small amount of boiling salted water, cover, and simmer until tender (about 12 to 15 minutes); drain.

## Seasoning and Serving

Cooked chayote tastes something like summer squash, but it has firmer flesh that is juicy and crisp like water chestnut. Season it with butter alone or butter to which a dash of thyme, marjoram, oregano, or basil leaves has been added. You can sprinkle hot, cooked, buttered chayote with a little crumbled cooked bacon, sliced green onion, shredded Cheddar cheese, or grated Parmesan cheese before serving.

### Chayote Appetizer

Served raw, chayote makes an interesting vegetable appetizer. Cut it in half lengthwise, lift out and discard the seed, and peel with a vegetable peeler. Thinly slice the halves and arrange slices on a serving plate with some lime wedges and a small dish of salt. To eat, dip slices in salt and sprinkle with lime juice.

---

### Butter-Steamed Chayote

Rinse 1 medium-sized (about 14 oz.) chayote, peel it, then slice crosswise through seed and all to make 1/8-inch-thick slices; cut the slices in half (you should have about 3 cups). Melt 2 tablespoons butter or margarine in a wide frying pan (one with a tight-fitting lid) over medium heat. Add sliced chayote, 2/3 cup water, 1/4 teaspoon salt, and 1/8 teaspoon pepper. Add 1/4 teaspoon oregano or thyme leaves, if desired.

Stir mixture, cover tightly, and cook for 8 to 10 minutes or until chayote is just tender and liquid is absorbed. Stir in 2 tablespoons chopped parsley and turn into a serving dish. Serve at once. Makes 3 to 4 servings.

# Marinated Chayote Salad

3 chayotes, sliced or diced
½ pound fresh green beans
  Boiling salted water
½ cup salad oil
¼ cup vinegar
1 teaspoon lime juice
½ teaspoon *each* salt and sugar
¼ teaspoon tarragon leaves, crumbled
  Dash pepper
2 medium-sized tomatoes cut in wedges, or several cherry
    tomatoes
  Crisp lettuce leaves

To cook sliced or diced chayotes, follow directions under *cutting and cooking,* page 37; drain and cool. Also cook beans in boiling salted water until just tender; drain and cool. Mix together the salad oil, vinegar, lime juice, salt, sugar, tarragon and pepper. Add the chayote and beans to the dressing, and marinate in the refrigerator.

To serve, arrange the pieces of drained chayote, beans, and tomatoes on crisp lettuce leaves. Makes about 6 servings.

# Chayote and Tomato Salad

2 medium-sized (about 14 oz. *each*) chayotes, peeled and
    cut in half lengthwise
3 tablespoons olive or salad oil
2 teaspoons lime juice
3 tablespoons white wine vinegar or white vinegar
⅛ teaspoon pepper
¼ teaspoon *each* salt, sugar, and basil leaves, crumbled
⅓ cup *each* finely chopped green pepper and green onion
  Lettuce leaves
2 medium-sized tomatoes, peeled and cut in thin wedges

Discard seed from chayote and thinly slice halves, then cut slices into thirds. Drop into 2 quarts boiling salted water and cook for 1 minute. Drain and plunge into a large quantity of cold water; drain again.

In a bowl, mix the oil, lime juice, vinegar, pepper, salt, sugar, and basil. Add the chayote, green pepper, and onion. Cover and chill for at least 1 hour.

To serve, line a salad bowl with lettuce leaves, fill with chayote mixture, and garnish with tomato wedges. Makes 6 servings.

# Panama-Style Stuffed Chayotes

2 medium-sized (about 14 oz. *each*) chayotes, unpeeled
  Salt
1 tablespoon butter or margarine
½ chopped onion
¾ pound ground lamb
2 tablespoons fine dry bread crumbs
1 teaspoon dried mint leaves, crumbled
½ cup seedless raisins
¼ cup catsup
2 eggs
1 clove garlic, minced or mashed
¼ cup shredded fresh Parmesan cheese

Scrub chayotes well, cut in half lengthwise through seed. Pre-cook chayote halves as directed under *cutting and cooking,* page 37; drain and cool.

When cool enough to handle, use a spoon to scoop out pulp and seeds, leaving shells at least 1/4 inch thick; sprinkle lightly with salt. Chop pulp and seeds; set aside.

In a frying pan over medium heat, melt butter and put in onion. Cook, stirring occasionally, until golden

(6 to 8 minutes). Remove from heat and lightly mix in the chayote pulp, lamb, bread crumbs, mint, raisins, catsup, eggs, and garlic.

Mound filling into chayote shells, sprinkle Parmesan evenly over the top, and arrange in a shallow baking pan; pour in 1/4 inch water. Bake, uncovered, in a 375° oven until top is lightly browned and mixture heated through (30 to 35 minutes). Serve hot. Makes 4 servings.

# CORN

## Selection and Storing

**Selection:** Choose corn ears with green, fresh husks having silk-ends that are free from decay or worm injury. Cobs should be filled with plump, milky kernels.

**Buying:** Allow 1 to 2 ears corn for each serving. Two ears will yield about 1 cup fresh corn kernels.

**Storage:** Keep in husks, unwashed, until ready to use. Store in the refrigerator. Use as soon as possible.

## Cutting and Cooking

**To cook whole corn on the cob,** remove husks and silk just before cooking; a brush will help remove the silk. Slip ears of corn into a large kettle of unsalted boiling water to cover. Cook, covered, for 3 to 5 minutes; drain.

**When recipe calls for cooked corn,** follow directions above for cooking whole corn on the cob. When cool enough to handle, use a sharp knife to cut corn off cob, leaving the rough kernel bases attached to the cob.

**For uncooked whole kernel corn,** cut raw corn kernels off cob the same as for cooked corn to use as directed in recipes.

**For creamy uncooked corn,** first draw a sharp knife through the center of each row of corn kernels. Cut corn off the cob, then scrape with the knife to remove all the corn pulp and milk.

## Seasoning Corn on the Cob

Fresh young corn, cooked briefly as directed above, needs nothing more than salt, pepper, and butter for seasoning. But for variety try blending some chile powder, oregano leaves, or cumin seed with soft butter. For guest meals you might mix salt, pepper, and another seasoning, if desired, into soft butter, then form it into balls and chill. Or try mixing salt, pepper, and chile powder, then putting it in a salt shaker to pass at the table with butter. The following are other simple ways to cook and serve corn on the cob.

### Savory Baked Corn on the Cob

Remove husks and silk from 6 medium-sized ears corn. In a bowl combine 3 tablespoons mayonnaise with 1 clove garlic (minced or mashed), 1 teaspoon olive oil, 1/2 teaspoon smoke-flavored salt, and dash pepper. Spread each ear with mayonnaise mixture, then wrap tightly in foil. Roast in a 325° oven for about 30 minutes, or until tender.

### Barbecued Corn on the Cob

Blend 1/4 cup melted butter with 1 1/2 tablespoons soy sauce. Remove husks and silk from 6 medium-sized ears of corn. Place each ear on a sheet of heavy foil. Pour about 1 tablespoon of the butter mixture over each. Wrap securely and place in the barbecue coals for about 15 minutes, turning several times.

### Corn Grilled with Bacon

Cook whole corn on the cob as directed under *cutting and cooking* except boil only about 2 minutes. Spiral 1 or 2 thin slices bacon around each ear, fastening it at each end with wooden picks. Grill over charcoal or in the broiler until bacon is crisp and brown, turning frequently.

## Seasoning Cut Corn

These quick and easy corn recipes call for either cooked or uncooked corn, cut off the cob. Follow directions under *cutting and cooking* suggestions for preparing the corn—or leftover corn can be cut off the cob and used for cooked corn; frozen corn can substitute for uncooked corn kernels.

### Creamy Corn

In a frying pan melt 2 tablespoons butter or margarine. Add 2 tablespoons all-purpose flour and cook, stirring, until bubbly. Remove from heat and gradually stir in 1 cup light cream (half-and-half) or milk; cook, stirring, until thickened. Add 3 cups cooked corn kernels, 1 tablespoon sugar, 1 teaspoon seasoned salt, and a dash *each* pepper and ground nutmeg. Cook until heated through. Makes 4 to 6 servings.

### Succotash

In a saucepan melt 1/4 cup butter or margarine. Stir in about 2 cups cooked corn kernels, 1 1/2 cups cooked lima beans (or 1 3/4 cups cut green beans), 1/4 cup finely chopped green onion (including part of tops), and 1/2 cup whipping cream. Simmer until slightly thickened and heated through. Season to taste with seasoned salt and pepper. Makes 4 to 6 servings.

### Scalloped Corn

In a frying pan sauté 1/2 cup chopped green onions, including part of the tops, in 1/4 cup butter or margarine. Stir in 1 cup toasted bread cubes, 3 cups cooked corn kernels, 1/2 teaspoon salt, and 1/4 teaspoon thyme leaves; cook, stirring, until heated through. Makes 4 to 6 servings.

### Chinese-Style Corn With Ham

In a large frying pan or wok, sauté 1/4 pound *each* sliced mushrooms (about 1 1/2 cups) and ground cooked ham in 3 tablespoons butter or margarine for 2 minutes, stirring constantly. Then add 1 package (10 oz.) frozen Chinese edible pea pods, thawed, 1 can (5 oz.) water chestnuts, (drained and sliced), 3 cups cooked corn kernels,

2 teaspoons soy sauce, and salt to taste. Cook, stirring, until vegetables are just tender and thoroughly heated. Makes about 6 servings.

### Sesame Sautéed Corn

In a frying pan sauté 1 clove garlic (minced) and 1/3 cup sesame seed in 2 tablespoons butter or margarine until the seeds are lightly browned. Add 3 cups cooked corn kernels, 1/4 cup chopped parsley, 3 tablespoons grated Parmesan cheese, 1/4 teaspoon salt, and dash pepper. Cook, stirring, until the mixture is heated through. Makes 4 to 6 servings.

---

### Butter-Steamed Corn

Melt 2 tablespoons butter in a wide electric frying pan set on high heat or a wide frying pan over direct high heat. Add 4 cups uncooked corn kernels (or frozen whole kernel corn, thawed) and 4 tablespoons water. Cover and cook, stirring frequently, for 4 minutes (3 minutes for frozen corn). Salt to taste. Makes 4 to 5 servings.

### Cream-Glazed Corn With Coriander

Prepare corn as directed above for Butter-Steamed Corn. Add 3/4 teaspoon ground coriander or 1/4 teaspoon ground cloves along with the vegetable. After 4 minutes, remove cover and add 4 tablespoons whipping cream, stirring until liquid is almost gone. Makes 4 to 5 servings.

# Fresh Corn Fritters

5 tablespoons pancake mix
¼ teaspoon salt
⅛ teaspoon pepper
½ cup shredded Parmesan cheese
2 eggs, slightly beaten
2 cups uncooked corn kernels (see page 41)
2 tablespoons chopped pimiento
1 tablespoon chopped onion or green onion
1 to 2 tablespoons chopped canned California green chiles
    About 2 tablespoons butter or margarine

In a bowl combine the pancake mix, salt, pepper, and cheese. Blend in eggs, corn, pimiento, onion, and chiles. Heat the butter in a wide frying pan or griddle. Drop vegetable mixture (about 2 tablespoons for each fritter) onto the pan and flatten with the back of a spoon; cook over medium heat until lightly browned on bottom (about 3 minutes); turn, and cook about 3 minutes more. Serve immediately. Makes 8 fritters or 4 servings.

# Fresh Corn Soufflé

4 tablespoons butter or margarine
2 tablespoons chopped onion
5 tablespoons all-purpose flour
1 teaspoon salt
    Dash ground pepper
¾ cup milk
2 cups uncooked corn kernels (about 4 to 6 ears)
5 eggs, separated

Melt butter in a pan; add onion and cook until limp. Stir in flour, salt, and pepper, and cook until bubbly. Remove from heat, gradually add milk; return to heat and cook, stirring until thickened. Stir in corn and cook 1 minute longer. Remove from heat and beat in egg yolks. If necessary, return to heat and stir a few seconds to thicken again; set off heat. Whip egg whites until they hold short, distinct, moist peaks. Fold half the whites thoroughly into sauce; fold in remaining whites as thoroughly as you like.

Pour into a well buttered 1 1/2-quart soufflé dish or about 6 individual soufflé dishes. Bake in a 375° oven 35 minutes (15 to 20 minutes for small soufflés). Makes 6 servings.

# Fresh Corn Omelet

6 eggs, separated
2 cups cooked corn kernels (directions page 41)
1 teaspoon salt
    Dash pepper
2 tablespoons chopped parsley or chopped green onions, including part of tops
½ teaspoon ground cumin
4 tablespoons melted butter or margarine

Combine the egg yolks with corn, salt, pepper, parsley or onion, cumin, and 2 tablespoons of the butter.

Heat remaining 2 tablespoons butter in a 10-inch oven-proof frying pan. Beat egg whites until stiff but not dry; fold corn mixture into egg whites and spoon the egg mixture into the warm frying pan. Cook over low heat until egg mixture is set around edge of pan and golden on the bottom. Place in a 350° oven and bake 15 to 20 minutes or just until set. Serve immediately. Makes 4 to 6 servings.

# Corn Chowder

8 slices bacon
1 large onion, diced
1 quart water
4 cups diced peeled potatoes
2 cans (1 lb. *each*) cream-style corn
1 quart milk (or 1 pint *each* milk and light cream)
    About 3 teaspoons salt
    About ½ teaspoon seasoned pepper
2 tablespoons butter or margarine

In a Dutch oven (at least 5-quart size) cook bacon until crisp; remove bacon from pan, drain, then cover and refrigerate until needed. Discard all but 2 tablespoons bacon drippings. Add onion to reserved drippings in pan and sauté until onion is soft. Add water, potatoes, and corn. Cover and simmer for about 20 minutes. Add milk and season to taste with salt and seasoned pepper; cool and refrigerate if made ahead.

Reheat slowly, stirring constantly; add butter. Crumble bacon and recrisp, uncovered, in a 350° oven for about 4 minutes. To serve, sprinkle bacon over soup. Makes about 8 servings.

# Fresh Corn Appetizer Dip

½ teaspoon salt
1 package (3 oz.) cream cheese, softened
1 tablespoon lemon juice
3 tablespoons whipping cream or sour cream
1 small clove garlic, mashed
1 teaspoon oregano leaves, crumbled
2 tablespoons *each* chopped pimiento and chopped green onion, including parts of tops
1½ cups cooked corn kernels (see page 41)

In a small bowl blend together cream cheese, lemon juice, cream, garlic, oregano, pimiento and green onion. Add corn and stir to blend. Cover and refrigerate for 2 hours. To serve, spoon into shallow serving dish. Serve with corn chips or any fairly sturdy chip or cracker. Makes about 2 cups dip.

# Mexican Cornbread

2 eggs
¼ cup salad oil
1 to 4 canned California green chiles, seeded and finely chopped
1 small can (about 9 oz.) cream-style corn
½ cup sour cream
1 cup yellow cornmeal
½ teaspoon salt
2 teaspoons baking powder
2 cups (about 8 oz.) shredded sharp Cheddar cheese

In a large bowl, beat eggs and salad oil until well blended. Add chiles, corn, sour cream, cornmeal, salt, baking powder, and 1 1/2 cups of the cheese to the egg mixture;

stir until thoroughly blended. Pour into a greased 8 or 9-inch round or square pan. Sprinkle remaining 1/2 cup cheese evenly over the top. Bake in a 350° oven for 1 hour or until a pick inserted in the center comes out clean and crust is lightly browned.

Serve warm; or cool completely, wrap in heavy foil, and refrigerate or freeze. Reheat (wrapped in foil) in a 300° oven for 30 minutes if chilled, 1 hour if frozen. Makes 6 to 8 servings.

# Fresh Corn Salad

3 cups cooked corn kernels (about 6 to 7 ears)
1½ cups thinly sliced celery
1 red bell or green pepper
3 green onions, including part of tops, thinly sliced
2 tablespoons minced sweet pickle
⅓ cup mayonnaise
¼ cup sour cream
1 tablespoon vinegar
½ teaspoon *each* salt and Worcestershire
⅛ teaspoon pepper
½ teaspoon prepared Dijon-style mustard
¼ teaspoon basil leaves, crumbled

Prepare corn following directions under *cutting and cooking*, page 41. In a bowl combine corn, celery, half the bell pepper, chopped, onions, and pickle. Cover and chill. In a small bowl combine mayonnaise, sour cream, vinegar, salt, Worcestershire, pepper, mustard, and basil; cover and chill.

Before serving, combine vegetables with dressing; mix lightly. Serve on lettuce leaves garnished with remaining bell pepper cut in rings. Serves 6.

# Creamy Baked Corn

3 cups uncooked corn kernels (about 6 to 7 ears)
3 green onions, including part of tops, sliced
1 small green pepper, seeded and diced
2 tablespoons bottled Green Goddess dressing
¼ cup sour cream
1 teaspoon firmly packed brown sugar
1 tablespoon all-purpose flour
¼ teaspoon salt
⅛ teaspoon pepper
1 tablespoon butter or margarine
    Tomato Cheese Topping (recipe follows)

Prepare corn as directed under *cutting and cooking*, page 41. Put corn into a 1 1/2-quart casserole. Distribute onion and green pepper over corn. In a bowl, mix together until blended the dressing, sour cream, brown sugar, flour, salt, and pepper. Spoon evenly over vegetables. Dot with butter. Cover tightly with foil and bake in a 375° oven for 35 minutes. Stir before serving. Serves 6.

**Tomato Cheese Topping.** Arrange thin slices of 1 large tomato over top, then sprinkle with 1/4 cup shredded Parmesan; put back into oven, uncovered, for additional 10 minutes.

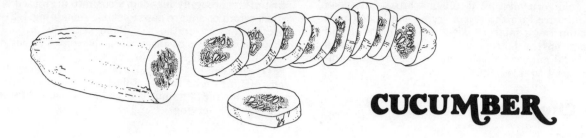

# CUCUMBER

## Selection and Storage

**Selection:** Choose firm, well-shaped cucumbers that are not too large in diameter. Exceptions are the pale green, coiled Armenian cucumbers and long, straight English cucumbers; both are grown big but are still good eating. Also the pale yellow-green lemon cucumbers, which are small and almost round.

**Buying:** Allow 1 regular cucumber for each 2 to 3 servings.

**Storage:** Store in the refrigerator, well wrapped to keep in moisture. They keep fresh several days.

## Cutting and Cooking

Wash when ready to use. **To serve raw,** remove skin with vegetable peeler, if desired, or score lengthwise with tines of fork. The thin wax coat often used on cucumbers is edible, but you may prefer to remove it. Cut in thick or thin slices or in sticks. If desired, let stand in salted ice water until crisp; drain.

**For cooked cucumbers,** peel and cut into thick slices; add to a saucepan with about 1 inch boiling salted water. Cook until just tender (about 10 to 15 minutes); drain.

## Serving Hot Cucumbers

Hot cucumber slices, cooked as directed above, can be seasoned with butter, salt, and pepper, and served hot. If desired, garnish with a little minced parsley or sliced green onion. Here are some other ways to cook cucumbers.

### Skewered Cucumbers with Mushrooms

Peel 2 medium-sized cucumbers and cut in bite-sized pieces. Trim 1/2 pound medium-sized mushrooms. Put vegetables in a bowl and pour over enough French dressing to coat all over. Marinate for several hours or overnight. Before serving, remove the vegetables from marinade and alternate on skewers with cherry tomatoes (about 1 basket). Place skewers about 5 inches from heat and broil or grill for about 5 minutes, turning until lightly browned all over. Makes 6 to 8 servings.

### Sautéed Cucumbers

Peel 2 or 3 cucumbers and slice enough to make 3 cups cucumber slices. In a frying pan sauté 1/4 cup chopped green onions (including part of tops) in 2 tablespoons butter or margarine. Add cucumbers, 1/4 teaspoon *each* salt and sugar, and dash pepper. Cook, stirring gently, for 2 to 3 minutes. Sprinkle generously with minced parsley or dill weed. Makes 6 servings.

### Sautéed Cucumbers with Tomatoes

Prepare sautéed cucumbers as directed above adding 1/4 teaspoon ground cumin with the salt, if desired. After cooking for 2 to 3 minutes, add 2 fresh tomatoes, peeled and cut in thin wedges; cook just until heated through. Instead of sprinkling with parsley or dill, spoon into a serving dish and sprinkle with 3 to 4 slices bacon, cooked and crumbled.

---

### Butter-Steamed Cucumbers

Melt 2 tablespoons butter in a wide electric frying pan set on high heat or in a frying pan over highest heat. Add 4 cups thinly sliced, peeled cucumbers (about 3 medium-sized), 1 tablespoon water, and 2 tablespoons vinegar or tarragon vinegar. Cover and cook, stirring frequently, for about 6 minutes. Season with salt and pepper. Sprinkle with paprika. Makes 5 to 6 servings.

### Cream-Glazed Cucumbers

Prepare cucumbers as directed above for butter-steaming. After cooking for 6 minutes, remove cover and add 3 tablespoons whipping cream and 1/2 teaspoon paprika. Stir and cook until liquid is almost gone. Sprinkle with paprika and minced parsley. Makes 5 to 6 servings.

## Marinated Cucumber Slices

This way with cucumbers is popular in many areas of the world, which has led to some interesting recipe variations. The long, thin English or Armenian cucumbers, which are almost free of seeds, are especially suitable for marinating. If you use regular cucumbers and they are fully mature, it is a good idea to cut them in half and scrape out any large seeds before slicing. Salting is an optional step in these recipes. Also peeling the cucumbers is optional. Use a vegetable slicer to make very thin slices.

**To pre-salt cucumbers,** put the slices in a bowl and sprinkle with salt, using about 1 teaspoon for each medium-sized regular cucumber. Mix with your hands, cover, and chill for 1/2 to 2 hours; rinse with cold water and drain well.

### Scandinavian-Style Marinated Cucumbers

Cut 2 medium-sized cucumbers (peeled if desired) in thin slices, and pre-salt them as directed above, if desired. Put the drained or freshly sliced cucumbers in a bowl and pour over this marinade: mix 1/4 cup white wine vinegar (or use tarragon or basil flavored vinegar), 1 teaspoon *each* salt and sugar, 1/2 teaspoon dill weed or minced parsley, and 1/4 teaspoon white pepper (optional). Cover and chill for at least 1 hour. Serve in the marinade or drain the cucumbers and arrange in serving dish, then spoon some of the marinade over to moisten. Makes 4 to 6 servings.

### Marinated Cucumbers with Shrimp

Follow recipe above for Scandinavian-Style Cucumbers. At serving time, lift cucumbers from marinade and arrange in overlapping slices around the edge of a serving plate. In the center make a mound of about 1/4 pound tiny cooked shrimp or 1 can (about 4 1/2 oz.) shrimp, drained and rinsed. Drizzle some of the cucumber marinade over shrimp to moisten.

### Oriental-Style Marinated Cucumbers

Slice 1 large (Armenian or English) or 2 medium-sized cucumbers in thin slices. Pre-salt as directed above, if desired. Put drained or freshly sliced cucumbers in a bowl and pour over 1/3 cup rice wine vinegar (or 1/4 cup white wine vinegar), 4 teaspoons sugar, 1 teaspoon salt, and 2 teaspoons grated fresh ginger (or use 1/2 teaspoon ground ginger). Cover and chill for at least 1 hour, or overnight. To serve, lift cucumbers out of marinade and serve in small dishes. Spoon a little of the marinade over top to moisten. Makes 4 to 6 servings.

## Cucumber Appetizer

2 slender cucumbers
½ cup cottage cheese
2 green onions, minced
¼ cup finely minced celery
½ teaspoon dill weed, tarragon or marjoram leaves
¼ teaspoon salt
   Pepper to taste
   Minced parsley
   Watercress (optional)

Peel cucumbers with a garnish knife or run fork tines down the sides of the peeled cucumber. Cut in 1-inch slices, and scoop out part of the insides to form a cup. Combine the cottage cheese, onions, celery, dill, salt, and pepper and fill cucumber cups with mixture. Sprinkle with minced parsley and serve garnished with watercress, if desired. Makes 8 to 12 servings.

## Molded Fresh Cucumber Salad

2 envelopes unflavored gelatin
½ cup cold water
4 medium-sized cucumbers
¾ cup *each* sour cream and mayonnaise
2½ tablespoons prepared horseradish
2 tablespoons grated onion
1 teaspoon salt
¼ teaspoon white pepper
1 cup whipping cream, whipped
   Thinly sliced cucumbers and watercress for garnish

Soften gelatin in cold water and place over hot water to dissolve. Pare cucumbers, cut them in half lengthwise, and remove seeds. Chop and whirl in blender to purée. Measure 3 cups purée and combine with sour cream, mayonnaise, horseradish, grated onion, salt, and pepper. Stir in dissolved gelatin. Chill until thickened (about

45 minutes). Fold in whipped cream. Turn into lightly oiled 1 1/2-quart salad mold. Chill at least 4 hours or until set. Unmold and garnish with sliced cucumbers and watercress. Makes 8 servings.

## Mexican Cucumber-Orange Salad Tray

2 medium-sized cucumbers
⅓ cup white wine vinegar
½ cup olive oil or salad oil
½ teaspoon salt
1 large mild onion (red or white), cut in thin vertical slices
4 large oranges, peeled with a knife to remove white membrane and cut in crosswise slices
1 large avocado, peeled and sliced
   About 1 tablespoon lemon juice
   Butter lettuce leaves
   Orange-Chile Dressing (recipe follows)

With a vegetable parer peel strips lengthwise from each cucumber, making alternating patterns of green and white. Thinly slice cucumbers and place in a bowl, adding vinegar, oil, and salt; mix, cover, and chill at least 1 hour.

With a slotted spoon, remove cucumber from marinade (save marinade). On a large rimmed tray, arrange separately side by side the cucumbers, onions, oranges, and avocado; drizzle lemon juice over avocado to preserve color. Garnish tray with lettuce. If made ahead, cover with plastic film and chill up to 5 hours. At serving time, spoon Orange-Chile Dressing onto individual portions. Serves 8 to 10.

**Orange-Chile Dressing.** Add to reserved cucumber marinade 1 1/2 teaspoons grated orange peel, 1/2 teaspoon chile powder, and 1/4 teaspoon salt; do not refrigerate. Blend and pour into a small bowl.

# Cucumber Chile Dip

1 large cucumber, peeled and minced
½ teaspoon salt
1 small package (3 oz.) cream cheese, blended with 2
    tablespoons sour cream
2 tablespoons seeded and chopped canned California
    green chiles
    Salt to taste

In a bowl mix salt with cucumber and chill at least 1 hour to release liquid. Drain all liquid from cucumber and blend cucumber with cream cheese and sour cream; stir in chiles and salt to taste. Serve with fresh raw vegetables such as cauliflower, cherry tomatoes, green or red peppers or radishes.

# Cucumber Buttermilk Soup

1 large cucumber, peeled, seeded, and coarsely chopped
1 small green onion, finely chopped
2 tablespoons chopped green pepper
1 teaspoon salt
1 tablespoon lemon juice
½ teaspoon *each* Worcestershire, celery seed and dill weed
    Dash white pepper
1 quart buttermilk
1 teaspoon minced parsley

Combine the cucumber with green onion, pepper, salt, lemon juice, Worcestershire, celery seed, dill weed, and white pepper. Pour in buttermilk. Serve cold, garnished with minced parsley. Makes about 6 servings.

# EGGPLANT

## Selection and Storage

**Selection:** Choose firm, heavy eggplants with shiny, purple, uniformly smooth skin. Different varieties of eggplant, including small European and Oriental kinds, can be used interchangeably in most recipes.

**Buying:** Allow 1 medium-sized eggplant (1 1/4 to 1 1/2 lbs.) for 4 to 6 servings.

**Storage:** Keeps best at temperatures around 50° (such as in a cooler, basement, or wine cellar). If a suitable cool storage area isn't available, purchase close to serving time and refrigerate only briefly.

## Cooking Suggestions

Eggplant can be cooked in a variety of interesting ways: broiling, grilling over coals, sautéing, baking, and frying. These cooking techniques, which are different for eggplant than for any other vegetable, are all described in this section. The cooking method you choose will determine how the eggplant is cut up and the kinds of seasonings that may be added. Usually you merely rinse and trim the eggplant and use without peeling.

## Basic Grilled Eggplant

The eggplant develops rich flavor as it grills, seasoned only with salt and pepper—or you can introduce more flavors with the oil.

Wash, trim, and cut a medium-sized (about 1 1/4 lbs.) eggplant into 8 wedges or 3/4-inch-thick slices. Melt 1/3 cup butter or use 1/3 cup olive oil and brush all cut surfaces of eggplant with part of the butter or oil. Sprinkle with salt and pepper or seasoned salt and pepper to taste. Arrange eggplant pieces on the broiler pan or barbecue grill. Broil or grill 3 or 4 inches from heat, turning as is necessary until tender and well browned on all sides; brush with butter or oil several times. Total cooking time is about 10 minutes. Makes 4 to 6 servings.

### Italian Herb-Basted Eggplant

In a small pan combine 1/3 cup olive oil, 1/2 teaspoon salt or garlic salt, 1/2 teaspoon mixed Italian herbs (or marjoram), and 1/8 teaspoon pepper. Heat to blend flavors, then remove from heat. Follow directions for Basic Grilled Eggplant above, substituting this flavored oil for the melted butter or oil and seasoning.

### Lemon Butter-Basted Eggplant

Melt 1/4 cup (1/8 lb.) butter in a small saucepan; stir in 1 tablespoon lemon juice, 1/4 teaspoon salt, 1/4 teaspoon basil leaves, crumbled, and 1/8 teaspoon ground cinnamon. Follow directions for Basic Grilled Eggplant above, substituting this lemon butter for the melted butter or oil and seasonings.

### Soy-Basted Eggplant

Combine in a small pan 1/4 cup salad oil, 2 tablespoons soy sauce, 1 tablespoon Sherry or lemon juice, 1 clove garlic, crushed, and 1/4 teaspoon ground ginger; heat to blend flavors. Follow recipe for Basic Grilled Eggplant, substituting this soy baste for the melted butter or oil and seasoning. Makes 4 to 6 servings.

### Basic Eggplant Sauté

This method uses a minimum of fat and is quick to do. You'll need a heavy 10 or 12-inch frying pan that has a tight-fitting lid.

Wash, trim, and cut 1 small (about 1 lb.) eggplant into 8 wedges. In a heavy frying pan heat 2 tablespoons olive oil or salad oil over medium-high heat. Put in eggplant. As soon as all pieces are nicely brown on one side (about 2 minutes), turn them and immediately add 1/2 teaspoon *each* salt and crushed basil leaves and 3 tablespoons water. Cover pan and cook over medium heat; remove lid and check at 1 to 2 minute intervals, adding about 2 tablespoons water at a time as it is absorbed; shake pan or turn pieces as necessary until completely tender (about 12 minutes total cooking time). Sprinkle with more salt and pepper to taste, if needed. Makes about 4 servings.

#### Tomato-Eggplant Sauté

Follow recipe for Basic Eggplant Sauté. When eggplant pieces are almost tender (about 6 to 8 minutes), add 1 can (about 14 oz.) pear-shaped tomatoes and liquid with 1 teaspoon sugar and an additional 1/2 teaspoon basil leaves. Cover and cook about 2 minutes, then remove cover, increase heat, and continue cooking a few minutes to reduce liquid, if necessary. Makes about 4 to 6 servings.

### Pan-Fried Eggplant

This method of frying eggplant on top of the range results in a thin, crisp coating.

Wash and trim a medium-sized eggplant (about 1 1/4 lbs.); cut into fingers 3/4 inch thick. In a paper bag, combine 1/2 cup all-purpose flour, 1/4 cup cornstarch, 1 teaspoon salt, and 1/2 teaspoon pepper or seasoned pepper. Put eggplant pieces, a few at a time, into bag and shake to coat all over. Shake off excess flour.

In a large frying pan, heat just enough olive oil or salad oil to coat pan well. Place pan over medium heat until oil is heated. Put in just enough eggplant pieces at a time to cover pan bottom and cook until brown on one side. Turn as needed to brown all sides. Add more oil (about 1 tablespoon at a time), as needed to keep pan bottom lightly coated (about 1/3 cup total). When eggplant is tender and brown, remove and drain briefly. Salt and pepper to taste.

If desired, serve with a sauce such as canned marinara sauce that has been heated, a cold chile sauce, or the Creamy Sauce that follows. Makes 4 to 6 servings.

**Creamy Sauce for Fried Eggplant.** Combine in a bowl 1 cup sour cream or unflavored yogurt, 1 teaspoon *each* instant minced onion, dill weed, and salt (or garlic salt), 2 teaspoons lemon juice (omit with yogurt), and dash pepper. Stir until well blended, cover, and chill for at least 1 hour to blend the flavors. Makes about 1 cup.

### Oven-Fried Eggplant

This method of frying eggplant requires the least attention; it has a cracker-crumb crust.

Wash and trim a medium-sized eggplant. Cut in 1/2-inch-thick slices. Measure 1/4 cup mayonnaise; set aside. Crush salted soda crackers to make 1/2 cup crumbs, and mix with 1/4 cup grated Parmesan cheese in a shallow pan; set aside.

Lightly spread mayonnaise on one side of each eggplant slice, dredge in crumb mixture, then spread second side with mayonnaise and coat with crumb mixture. Arrange pieces in a single layer in an ungreased shallow baking pan. Put into a 425° oven until browned and tender (about 15 minutes), turning if needed to brown both sides evenly. Serve with any of the sauces suggested under Pan-Fried Eggplant, above, if desired. Makes 4 to 6 servings.

## Baked Stuffed Eggplant

2 small eggplants (1 lb. *each*)
  Salt
5 tablespoons olive oil
2 medium-sized onions, thinly sliced
3 large cloves garlic, minced or mashed
⅓ cup minced parsley
1 teaspoon sugar
1½ teaspoons salt
2 large tomatoes, peeled, seeded, and diced, *or* 1 can (1 lb.) whole tomatoes, drained and diced
  Unflavored yogurt (optional)

Cut eggplant in halves lengthwise and scoop out center seed sections, leaving a firm 1/2-inch-thick shell; reserve pulp. Sprinkle shells lightly with salt. Drizzle 2 tablespoons of the oil in a shallow baking pan; arrange eggplant in the pan.

In the remaining 3 tablespoons oil, sauté onion until limp. Stir in about 2 cups diced unseeded portions of the eggplant pulp, the garlic, parsley, sugar, 1 1/2 teaspoons salt, and tomatoes.

Distribute filling inside the eggplant shells. Cover the pan with foil and bake in a 350° oven until tender (about 1 1/2 hours). Cool in pan and serve at room temperature (bring to room temperature if cold); spoon pan juices over eggplant. Pass a bowl of unflavored yogurt to spoon on top, if desired. Makes 4 large servings of 1/2 eggplant each, or cut each in two for 8 smaller servings.

# Eggplant Slices with Pine Nuts

Here golden cream sauce blends with melted feta cheese and pine nuts to cover round eggplant slices.

1 large eggplant
6 tablespoons olive oil
3 tablespoons *each* butter and all-purpose flour
1 cup milk
2 egg yolks
3 ounces feta cheese, crumbled, or 1 package (3 oz.) cream cheese
2 tablespoons chopped parsley
⅓ cup pine nuts

Slice eggplant crosswise in 1-inch-thick rounds. Discard end slices. Brush both sides with oil and place in a shallow baking pan. Bake in a 450° oven for 15 minutes, turning once, or until tender.

In a small saucepan, melt butter and blend in flour. Gradually pour in milk; stirring constantly, cook until thickened, then 2 to 3 minutes longer. Beat egg yolks slightly; stir in the hot sauce, cheese, and parsley; cook 5 minutes longer, stirring. Spread sauce over eggplant slices and sprinkle with nuts. (Can refrigerate at this point.) Place in a 450° oven for 5 minutes (10 minutes if refrigerated), or until topping browns lightly. Makes 6 servings.

# Ratatouille
## (Rah-tah-too-yeh)

About ½ cup olive oil
2 large onions, sliced
2 large cloves garlic, minced or mashed
1 medium-sized eggplant, cut in ½-inch cubes
6 medium-sized zucchini, thickly sliced
2 green or red bell peppers, seeded and cut in chunks
About 2 teaspoons salt
1 teaspoon basil leaves
½ cup minced parsley
4 large tomatoes, cut in chunks
Parsley
Sliced tomato (optional)

Heat 1/4 cup of the oil in a large frying pan over high heat. Add onions and garlic and cook, stirring until onions are limp but not browned. Stir in the eggplant, zucchini, peppers, 2 teaspoons salt, basil, and minced parsley; add a little of the oil as needed to keep the vegetables from sticking. Cover pan and cook over moderate heat (about 30 minutes); stir occasionally, using a large spatula and turning the vegetables to help preserve their shape. If mixture becomes quite soupy during this time, remove cover to allow some of the moisture to escape.

Add the tomatoes to the vegetables in the pan and stir to blend. Add more oil if vegetables are sticking. Cover and cook over moderate heat for 15 minutes; stir occasionally. Again, if mixture becomes quite soupy during this period, remove cover and allow moisture to evaporate. Ratatouille should have a little free liquid but still be of a good spoon-and-serve consistency. Add more salt if desired. Serve hot or cover and chill to serve cold; reheat to serve. Garnish with parsley and tomato. Makes 8 to 10 servings.

**Ratatouille in the oven:** Using the vegetables and seasonings in the above recipe, layer all ingredients into a 6-quart casserole, pressing down to make fit. Drizzle only 4 tablespoons of the olive oil over top layer. Cover casserole and bake in a 350° oven 3 hours. Baste top occasionally with some of the liquid. Uncover during the last hour if quite soupy; this method of cooking makes a moist ratatouille. Mix gently and salt to taste. Serve hot, chilled, or reheated. Garnish with parsley and tomato.

# Athenian Moussaka

2 small (1 lb. *each*) eggplants
½ cup olive oil or salad oil
Meat Sauce (recipe follows)
Custard Topping (recipe follows)
½ cup shredded Parmesan cheese

Trim stem ends from eggplant and cut lengthwise in 1/4-inch-thick slices. Pour oil into two large baking pans and turn eggplant slices in it, coating both sides. Arrange in a single layer. Bake in a 425° oven for 30 minutes, turning occasionally, or until tender.

To assemble, arrange half eggplant in a 9 by 13-inch baking pan. Spoon on the meat sauce and top with remaining eggplant. Pour on custard topping and sprinkle with Parmesan. Bake, uncovered, in a 350° oven for 1 hour. Makes 12 servings.

**Meat Sauce.** Chop 2 medium-sized onions and sauté in 2 tablespoons salad oil. Add 2 1/2 pounds ground beef and cook, stirring, until crumbly. Add 2 teaspoons salt, 2 cans (6 oz. *each*) tomato paste, 1 1/4 cups dry red wine, 1/4 cup finely chopped parsley, 1 stick cinnamon, and 2 cloves garlic, minced. Cover and simmer 30 minutes. Uncover and remove cinnamon stick; cook down until liquid is evaporated. Stir in 3 tablespoons fine dry bread crumbs and 1 cup shredded Parmesan cheese.

**Custard Topping.** Melt 1/3 cup butter and blend in 1/2 cup all-purpose flour; cook, stirring, 2 minutes. Gradually stir in 1 quart milk and cook, stirring, until it boils and thickens. Add 1 teaspoon salt, 1/4 teaspoon ground nutmeg, and 1/2 cup shredded Parmesan cheese. Stir hot sauce into 6 slightly beaten eggs.

# Baked Eggplant Gratin

2 small eggplants (about 1 lb. *each*)
½ cup olive oil
Seasoned crumbs (recipe page 73)
4 to 5 tablespoons olive oil

Trim stem from eggplants and cut crosswise in 1-inch-thick slices. Pour the 1/2 cup olive oil in a 10 by 15-inch baking pan. Coat eggplant slices in oil; then pat 1 to 2 tablespoons of the seasoned crumbs on top of each slice; drizzle with the 4 to 5 tablespoons olive oil. Bake, uncovered, in a 375° oven for 1 1/2 hours. Cool to room temperature. Makes 6 to 8 servings.

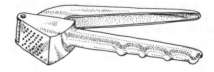

## Eggplant "Caviar" Appetizer

Offer this vegetable spread with warm rye bread and butter as a first course.

1 large eggplant
1 medium-sized onion, chopped
½ green pepper, chopped
2 tablespoons *each* olive oil and butter or margarine
1 large tomato, peeled and chopped
2 tablespoons lemon juice
1 tablespoon catsup
½ teaspoon Worcestershire
¼ teaspoon salt
   Dash pepper
   Dill weed, chopped chives, *or* chopped parsley for
    garnish
   Black olives for garnish

Place whole eggplant in a shallow, greased baking pan and bake in a 400° oven until soft (about 20 minutes). Remove and cool slightly. Then peel it and coarsely chop the pulp. In a frying pan, sauté the onion and green pepper in the olive oil and butter until vegetables are tender but not browned. Add the chopped eggplant, tomato, lemon juice, catsup, Worcestershire, salt, and pepper. Cook over high heat, stirring until liquid is cooked away; cool.

Spoon into shallow serving dish; sprinkle with dill weed, chopped chives, or chopped parsley. Garnish with several black olives; chill before serving. Makes about 4 servings.

## Eggplant Supper Soup

2 tablespoons *each* olive oil or salad oil and butter or
   margarine
1 medium-sized onion, chopped
1 pound lean ground beef
1 medium-sized eggplant, diced
1 clove garlic, minced or mashed
½ cup *each* chopped carrot and sliced celery
1 large can (1 lb. 12 oz.) pear shaped Italian-style
   tomatoes
2 cans (14 oz. *each*) beef broth
1 teaspoon *each* salt and sugar
½ teaspoon *each* pepper and ground nutmeg
½ cup salad macaroni
2 tablespoons minced parsley
   Grated Parmesan cheese

Heat salad oil and butter in a Dutch oven, add onion, and sauté until limp (about 3 minutes). Add the meat and stir over the heat until it loses its pinkness. Add eggplant, garlic, carrots, celery, tomatoes (break up with a fork), beef broth, salt, sugar, pepper, and nutmeg. Cover and simmer for about 30 minutes. Add macaroni and parsley and simmer about 10 minutes more, or until macaroni is tender. Serve in heated soup bowls. Pass cheese to sprinkle over. Makes 6 to 8 servings.

 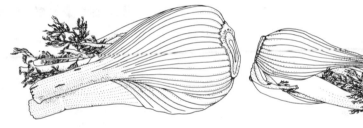

## Selection and Storage

**Selection:** Fennel is often called *finocchio* or *anise*. Choose fennel with fresh-looking bright green featherlike leaves and rigid, crisp stalks. The bulb should be pearly white.

**Buying:** Allow 1/2 to 1 medium-sized bulb for each cooked serving.

**Storage:** Do not rinse. Enclose in a plastic bag and store in refrigerator. Use within a few days.

## Cutting and Cooking

Both the celery-like bulb and the feathery green top can be cooked and served in a variety of ways. Wash and trim off hard outside stalks and coarse part of top. The fennel bulb can be cut lengthwise in halves or quarters, cut into strips, sliced, or diced.

**To cook fennel,** put the cut bulb into a small amount of boiling salted water. Cover and cook until just tender (about 8 to 10 minutes for halves or quarters, or 5 to 8 minutes for sliced or diced fennel); drain.

## Seasoning and Serving

Fennel can be served raw in sticks on a relish dish, used with a dip for an appetizer, or sliced or diced for salads. The raw pieces served with a blue-veined cheese is an especially good flavor combination.

Cooked fennel can be seasoned with butter or with melted butter, to which some of the green fennel top has been added.

### Fennel with Cheese

Cook fennel slices as directed under *cutting and cooking*; arrange cooked and drained fennel slices in a shallow baking dish. Pour over Mornay Sauce to cover (recipe page 6), using about 1/4 cup for each serving. Sprinkle lightly with grated Parmesan cheese. Bake in a 350° oven until bubbly and top is brown (about 20 minutes).

**Fennel in Cream**

Dice and cook fennel following the directions under *cutting and cooking*, page 49; drain. In a heavy saucepan, combine about 1/2 cup whipping cream with each 2 cups of diced, cooked fennel you use. Simmer the mixture rapidly, uncovered, over medium-low heat until cream is slightly reduced. Season with salt and pepper to taste.

# Baked Fennel

2 bulbs fennel, trimmed and cut into lengthwise quarters
    Boiling salted water
4 tablespoons melted butter or margarine
2 tablespoons fine dry bread crumbs
1 tablespoon grated Parmesan cheese
1 hard-cooked egg, chopped
1 tablespoon finely chopped fennel leaves
    Dash paprika

Cook according to directions under *cutting and cooking*, page 49. Arrange drained fennel in a well-greased, shallow baking dish. Spoon 2 tablespoons of the melted butter over fennel. Combine bread crumbs, cheese, egg, fennel leaves, and remaining 2 tablespoons melted butter; sprinkle over top. Add a sprinkling of paprika. Place in a 450° oven just until crumbs have browned (about 10 minutes). Makes about 4 servings.

# Green Beans with Fennel

1 bulb fennel, trimmed and diced
1 pound green beans, cut in 1-inch pieces
    Boiling salted water
4 tablespoons butter
1 tablespoon all-purpose flour
½ cup cold water
1 teaspoon lemon juice
    Few grains salt
2 teaspoons grated onion

Cook fennel according to directions under *cutting and cooking*, page 49. Cook beans in boiling salted water until tender-crisp (about 10 minutes). Drain and combine with fennel in a warm serving dish. In a small pan, melt 2 tablespoons of the butter, add flour and cook until bubbly. Stir in cold water and cook, stirring until thick. Stir in remaining 2 tablespoons butter, lemon juice, salt, and grated onion. Heat slightly. Pour over vegetables and serve. Makes 4 servings.

# Fennel and Tomato Salad

1 large bulb fennel, trimmed, cut in half lengthwise, and thinly sliced
½ head iceberg lettuce, thinly sliced
½ cup finely chopped fennel leaves
3 medium-sized tomatoes, peeled and sliced
¼ cup garlic wine vinegar
½ teaspoon salt
¼ teaspoon *each* garlic salt and pepper
½ cup olive oil or salad oil

Combine sliced fennel, lettuce, and fennel leaves in a bowl. Add sliced tomatoes. In a jar or bottle, combine the vinegar with the salt, garlic salt, pepper, and oil; shake to blend well. Pour over the salad, mix gently, and serve immediately. Makes 4 servings.

# Fennel in Tomato Sauce

1 green pepper, seeded and chopped
2 tablespoons olive oil or salad oil
1 teaspoon *each* paprika, salt, and sugar
1 can (8 oz.) tomato sauce
¼ cup water
2 large bulbs fennel, trimmed and sliced crosswise into ½-inch-thick slices
2 tablespoons grated Parmesan cheese
2 tablespoons finely chopped fennel leaves

In a large frying pan with a tight fitting lid, cook green pepper in oil until limp (about 5 minutes). Add paprika, salt, and sugar; cook, stirring, about 1 minute. Stir in tomato sauce and water; cook about 5 minutes. Carefully set fennel into tomato sauce. Cover and simmer gently until tender. Turn into a serving dish and sprinkle with grated cheese and fennel leaves. Makes 4 to 6 servings.

# Braised Fennel with Bacon

2 large bulbs fennel, trimmed and cut lengthwise into quarters
4 slices bacon, chopped
1 *each* medium-sized carrot and onion, chopped
1 cup regular strength beef broth

Cook fennel according to directions under *cutting and cooking*, page 49. In the bottom of a wide, heavy pan, evenly distribute the uncooked bacon, carrot, and onion. Top with the drained fennel. Add the beef broth; cover and simmer about 5 to 8 minutes or until flavors are well blended. Lift fennel from pan with a slotted spoon. Boil sauce rapidly until reduced by half. Pour over fennel and serve. Makes 4 to 6 servings.

# Marinated Fennel, Greek Style

2 cups water
½ cup olive oil
1 teaspoon salt
⅓ cup lemon juice
½ teaspoon coriander seed
¼ teaspoon whole black pepper
½ teaspoon thyme leaves
1 bay leaf
2 large bulbs fennel, trimmed and cut lengthwise in quarters

In a large pan, combine the water, olive oil, salt, lemon juice, and seasonings. Bring to a boil. Drop in the fennel and cook about 10 minutes until tender-crisp. Remove from heat; cover and chill in the cooking broth for several hours or overnight. Serve cold with a little of the cooking liquid as sauce. Makes 4 servings.

# GREENS

*Beet Tops, Belgian Endive,
Swiss Chard, Dandelion Greens,
Kale, Lettuce, Mustard Greens,
Spinach, Turnip Tops, Watercress*

## Selection and Storage

**Selection:** When buying greens such as *chard, kale, mustard, spinach, or turnip tops,* look for leaves that are fresh, tender, free from blemishes, and a good green color. They should have a minimum of yellowed leaves.

*In choosing lettuce,* look for signs of freshness: leaves of iceberg lettuce and romaine should be crisp; other varieties will be softer but should not look wilted. Heads of iceberg should be large, round, and solid but not hard. Also check lettuce for signs of tipburn (browning around margins of leaves).

**Buying:** For most greens you plan to cook, allow about 1 pound for each 3 to 4 servings; a pound of spinach serves 2 to 3.

*For salads,* a medium-sized head of most varieties of lettuce makes about 4 servings.

**Storage:** *Lettuce and leafy greens* need some moisture during storage but should not be wet. Rinse in cool, running water and drain thoroughly; handle carefully, for the leaves bruise easily. Store in the crisper section of the refrigerator and use as soon as possible.

*When you want greens for salad* to be cold, crisp, and dry, follow this method: cut off stem end or cut out core, gently separate leaves, and wash well. Stand up or arrange leaves so water can drain off quickly. When nearly dry, wrap loosely but completely in a dampened cloth or paper towels and store in the crisper of the refrigerator (or seal towel-wrapped greens in a large plastic bag).

*Watercress* should be rinsed and then put in a container with stems in water, a plastic bag slipped over the tops, the bag secured to container, and refrigerated.

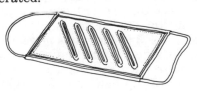

## Preparation and Cooking

When ready to cook, cut off roots; remove any badly damaged or yellowed leaves, then wash thoroughly for leafy greens provide many niches where bits of soil and sand may lodge. The best way to wash greens is to swish them vigorously about in a large quantity of cool water; lift from water, then repeat the washing process as many times as necessary. Let greens drain for a few minutes in a colander or on a rack.

**To cook leafy greens,** place them in a large pan with a tight fitting lid. Usually the liquid that clings to the leaves of such tender greens as spinach and chard is sufficient liquid for cooking; you may need to add a little liquid to greens that require longer cooking, such as kale or mustard greens. Cover and cook just until wilted. Recipes for cooking individual greens follow. Many greens are also interestingly served as a soup.

---

### Spring Greens Soup

In a saucepan, combine 2 cans (14 oz. *each*) regular strength chicken broth with 1/4 teaspoon salt, 3/4 teaspoon sugar, 2 teaspoons soy sauce, 2 to 4 thin slices fresh ginger root (optional), and 1/2 cup water. Simmer for about 15 minutes.

Meanwhile wash and prepare any of these greens: 1 small head chicory, 1 bunch mustard greens, 1 small bunch spinach, or 1 large bunch watercress. Discard tough ends and stems. Coarsely chop or break the leaves into bite-size pieces and loosely pack into a cup (you should have 2 to 3 cups).

Bring soup to a full boil and add the prepared greens selected; cover, reduce heat, and simmer 2 to 5 minutes, or until leaves are tender (exact time depends on your choice of greens). Add about 2 tablespoons thinly sliced green onion during last 2 minutes. Serve at once in mugs or soup bowls. Makes 4 to 6 servings.

# BEET TOPS

Tender, young leaves trimmed from beets have mildly sour and salty flavor. Discard tough stems and wash well as directed on page 51.

### Butter-Steamed Beet Tops

Melt 2 tablespoons butter or margarine in a wide heavy frying pan or electric frying pan. Add 4 cups cleaned, slightly damp, chopped beet tops, and 3 tablespoons regular strength beef broth or water; stir well. Cover and cook over medium-low heat for about 7 minutes, stirring occasionally. Season with 1/4 teaspoon basil leaves, if desired, and salt and pepper to taste. Makes about 3 generous servings.

### Beet Tops in Cream

In a wide, heavy frying pan or electric frying pan, cook 4 slices diced bacon until browned; drain off all but 2 tablespoons drippings. Add 4 cups washed, slightly damp, chopped beet tops, 1/4 cup regular strength beef broth, and 1/4 cup half-and-half. Cover and cook over medium heat for about 5 minutes, stirring occasionally. Add 2 to 3 teaspoons tarragon vinegar and simmer rapidly, uncovered, until liquid is slightly reduced. Season with salt and pepper. Makes about 3 servings.

# BELGIAN ENDIVE

The slender white heads of Belgian endive (also called French endive or Witloof chicory) have a pleasantly bitter taste. Simply rinse before using. As a cooked vegetable, it is usually left whole or split in half lengthwise. It may be halved, separated in individual leaves, or sliced for salads.

### Butter-Braised Endive

Wash 1 pound Belgian endive (about 12 endives) and arrange in a frying pan (one with a lid). Cover with boiling salted water, bring to boiling, and boil 5 minutes, drain well. To the pan, add 2 tablespoons butter or margarine, 1 tablespoon lemon juice, and salt and pepper to taste. Cover and continue to cook over low heat until endive is tender (about 10 minutes). Makes 3 to 4 servings.

### Braised Endive with Lemon Sauce

Wash 3/4 pound Belgian endive (about 8 endives) and arrange in a frying pan (one with a lid). Cover with boiling salted water, return to boil, and boil 5 minutes; drain thoroughly. Add 1/4 cup (1/8 lb.) butter or margarine to the pan and sauté endive slowly on all sides for about 5 minutes. Pour in 1 cup regular strength chicken broth and simmer, uncovered, for about 20 minutes. Remove endive with a slotted spoon and arrange in a serving dish; keep warm (reserve broth). In a bowl beat 2 eggs with 3 tablespoons lemon juice; gradually beat in reserved broth. Pour sauce over endive. Makes 4 servings.

### Belgian Endive Salad, Curry-Nut Dressing

For the dressing, combine in a jar 1 teaspoon prepared Dijon-style mustard, 1 tablespoon lemon juice, 3 tablespoons salad oil, 1/3 teaspoon ground cinnamon, 1/2 teaspoon curry powder, 1/4 teaspoon salt, and 1 1/2 tablespoons chopped filberts or toasted almonds. Shake to blend, then let stand for about 1 hour. Line 4 to 6 individual salad plates with crisp romaine leaves (about 1 head romaine). Wash 6 to 9 Belgian endives (about 3/4 lb.) and cut in half lengthwise; arrange about 3 halves on the romaine for each plate. Shake dressing again and pour over each plate. Makes 4 to 6 servings.

# CHARD, SWISS

Swiss chard has dark green leaves with heavy, smooth white stems. Usually the heavy stem is cut out of each leaf, sliced, and cooked longer than the thin green leaf. Or the white and green parts may be cooked and served separately.

The Oriental vegetable, *chard cabbage* (Chinese call it *baak choy*), is like Swiss chard except that its thick-stemmed leaves are in heads. When the head is broken apart, it can be cooked in the same ways as Swiss chard.

### Butter-Steamed Swiss Chard

Melt 2 tablespoons butter or margarine in a wide frying pan or electric frying pan. Add thinly sliced stems from 1 pound washed Swiss chard. Cover and cook, stirring occasionally, for 3 to 4 minutes. Stir in the chopped leaves. Cook, covered, for 3 to 4 minutes more. Season to taste with salt and pepper. Makes 3 to 4 servings.

### Swiss Chard Soup

In a wide, heavy pan melt 2 tablespoons butter or margarine. Add chopped, heavy stems from 1 pound washed Swiss chard; cook, covered, stirring occasionally for 3 to 4 minutes. Stir in the chopped leaves and cook 3 to 4 minutes more. Sprinkle with 2 tablespoons flour and stir until blended. Gradually blend in 1 1/2 cups regular strength chicken broth and 1/2 cup half-and-half (light cream) or milk; cook and stir until slightly thickened. Season to taste with salt and pepper. (Whirl soup in blender if a smooth consistency is desired.) Makes 4 to 6 servings.

### Swiss Chard with Bacon

Wash 1 1/2 pounds Swiss chard. With a knife, cut the white center stalks of each leaf away from leafy green parts. Cut the white stalks in 1/4 inch slices, then slice the green leafy parts; keep separate. In a large frying pan or Dutch oven, cook 6 slices bacon until crisp; remove and drain. Return 2 tablespoons of the bacon drippings to the frying pan. Add to pan 1 medium-sized onion (chopped), white part of chard, and 2 tablespoons water; sauté for 3 minutes, then cover and cook until tender (about 10 minutes). Add green part of chard and 2 tablespoons water to pan; cover and cook until greens have wilted (about 7 minutes). Remove from heat, crumble in bacon, and mix lightly. Turn into an ovenproof serving dish and sprinkle with about 3/4 cup shredded Swiss or jack cheese. Place under broiler until cheese melts. Makes about 6 servings.

# DANDELION GREENS

The long slender leaves of dandelion are grown commercially for produce markets, or you might gather them in fields when they are young and tender. They have a snappy, tart bitter taste.

### Butter-Steamed Dandelion Greens with Ham

In a wide, heavy frying pan or electric frying pan, lightly brown 1 tablespoon butter or margarine. Stir in 4 cups washed, chopped dandelion greens and 2 tablespoons beef or chicken broth (or water); stir well. Cover and cook over medium-low heat for about 10 minutes, stirring occasionally. Add 2 tablespoons whipping cream and 1/4 cup diced cooked ham. Simmer rapidly until liquid is almost gone. Season with salt and pepper. Makes 3 to 4 servings.

### Blanched Creamed Dandelion Greens

Immerse 4 cups washed dandelion greens in boiling salted water. Let cook until water returns to a vigorous boil. Drain greens thoroughly and chop. In a frying pan, cook until soft 1/2 cup diced onion in 2 tablespoons butter or margarine. Add greens and blend well. Stir in 2 tablespoons sour cream and 1 teaspoon hot prepared mustard. Season with salt and pepper. Makes 3 servings.

### Dandelion Greens and Artichoke Salad

Trim ends from about 3/4 pound dandelion greens and coarsely chop enough leaves to make 3 cups; turn into a salad bowl. Tear enough romaine into bite-sized pieces to make 3 cups. Place in bowl. Add 1 bunch radishes, trimmed and thinly sliced.

For dressing, drain marinade from 2 jars (6 oz. *each*) marinated artichoke hearts into a bowl and add 2 tablespoons red wine vinegar, 1/4 teaspoon Dijon-style mustard, 1/4 teaspoon salt, and 1 clove garlic, minced; mix well. Pour dressing over greens and mix lightly. Arrange artichoke hearts on top. Serves 6.

# KALE

Most varieties of kale in produce markets have large curly leaves. The robust flavor of this vegetable complements foods like sausage and bacon. Remove the tough stems and chop the leaves before cooking.

### Butter-Steamed Kale

In a wide, heavy frying pan or electric frying pan, lightly brown 3 tablespoons butter or margarine or heat 3 tablespoons meat drippings (such as you might have from cooking a roast). Stir in 1 bunch (about 10 cups) washed, chopped kale, 1/4 cup beef or chicken broth, and 1 teaspoon dill weed. Cover and cook over medium-low heat for about 15 minutes, stirring occasionally. Makes 6 to 8 servings.

### Butter-Steamed Creamed Kale

Prepare Butter-Steamed Kale following directions above. Just before serving, stir in 1/2 cup sour cream and season with salt and pepper. Makes 6 to 7 servings.

### Butter-Steamed Kale with Sausage

Break apart 1 pound bulk pork sausage and brown in a wide, heavy frying pan or electric frying pan. Drain off all but 2 tablespoons drippings. Stir in 1 bunch (about 10 cups) chopped kale and 1/2 cup regular strength beef broth. Cover and cook over medium-low heat for about 15 minutes, adding more broth if needed. Serves 6 to 8.

# LETTUCE

Iceberg, or head lettuce, as well as romaine, escarole, and other varieties of leaf lettuce can be served in more ways than salads. They can be quickly cooked like spinach or made into soups.

### Butter-Steamed Lettuce

Heat 2 tablespoons butter or margarine in a wide, heavy frying pan or electric frying pan until lightly browned. Add 6 cups shredded head lettuce, stirring well. Cover and let cook over medium heat, stirring occasionally for 2 to 3 minutes or until lettuce is wilted. Season with salt and pepper to taste. Makes 4 servings.

### Butter-Steamed Lettuce with Nuts

Brown 2 tablespoons slivered or diced almonds in 2 tablespoons melted butter. Add 6 cups shredded lettuce and cook as directed above for Butter-Steamed Lettuce. Makes 4 servings.

### Creamed Lettuce

Cook lettuce as directed above for Butter-Steamed Lettuce. Just before serving, add 3 to 4 tablespoons whipping cream and simmer rapidly, uncovered, until slightly reduced. Makes 4 servings.

### Lettuce Soup

Cook 6 cups shredded or chopped head lettuce in 2 tablespoons butter or margarine as directed for Butter-Steamed Lettuce. Remove from heat and whirl smooth in a blender with 1 tablespoon chopped chives, 3/4 cup regular strength chicken broth, and 1/3 cup half-and-half (light cream). Return to pan and heat to simmering or chill and serve cold. Makes 4 to 5 servings.

# MUSTARD GREENS

The young leaves of this plant have frilly leaves and a peppery-bitter flavor.

### Blanched Mustard Greens

Immerse 4 cups mustard greens in boiling salted water; when water returns to boil, drain greens thoroughly and chop. In a saucepan, cook until soft 1/2 cup diced onion in 2 tablespoons butter or margarine. Stir in the greens and 1/4 cup diced cooked ham; heat through. Makes 4 servings.

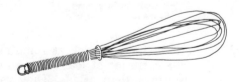

### Mustard Greens with Bacon

In a wide heavy frying pan or electric frying pan, cook 4 slices diced bacon until browned. Drain off all but 2 tablespoons drippings; stir in 4 cups chopped mustard greens and 2 tablespoons regular strength beef broth. Cover and cook over medium heat, stirring occasionally, for about 10 minutes. If desired, stir in 2 tablespoons shredded sharp Cheddar cheese; heat just until melted. Serves 4.

# SPINACH

A versatile green spinach is most widely used as a cooked vegetable. When eggs, fish, or meats are served on a bed of well-seasoned cooked spinach, the term *à la Florentine* is used to describe such dishes. Raw spinach leaves in salads contribute a clean, greens taste.

### Florentine-Style Dishes

First prepare spinach mixture: Melt 2 tablespoons butter in a wide frying pan over highest heat. Add 4 cups well drained, chopped, cooked fresh spinach (or use frozen chopped spinach, thawed and drained). Cook over high heat, stirring, until liquid is evaporated. Sprinkle with 2 teaspoons flour and mix well. Add 1/2 cup whipping cream and continue to cook and stir over high heat until there is no free flowing liquid. Season to taste with salt. Spread over bottom of a shallow casserole in a layer 3/8 to 1/2 inch deep, building spinach up sides of dish forming a rim. Or using 1/2 cup spinach for each serving, you can make up servings in individual ramekins.

Arrange hot cooked food (suggestions follow) evenly over spinach. For each serving allow about 1/4 pound boneless meat, 1/3 pound boneless fish, or 2 eggs. Spoon about 2 cups Mornay Sauce (recipe page 6) over the food. Or use about 1/4 cup Mornay Sauce in individual ramekins. If desired, sprinkle a little grated Parmesan cheese over the sauce. Put into a 500° oven and bake for 4 to 5 minutes (8 to 12 minutes if made ahead and cold); if desired, brown lightly under broiler. Makes 8 servings.

**Foods to use in Florentine dishes.** Hot poached eggs, cold hard-cooked eggs in halves or quarters; hot poached sole fillets, or oysters; canned tuna, oysters, salmon, lobster, or crab that has been drained; cold fresh lobster or crab; hot or cold cooked shrimp; hot boned chicken breasts either poached or sautéed in butter; cold sliced chicken or turkey; hot thin slices tender veal that has been browned in butter; hot broiled or sautéed boned lamb chops.

### Creamed Spinach

Cut off and discard coarse stems from 2 pounds spinach; wash thoroughly. Cook spinach, with the moisture that clings to leaves, in a large pan until tender (5 to 7 minutes). Drain well, reserving the liquid. Chop cooked spinach well or whirl in a blender.

In a saucepan, melt 3 tablespoons butter or margarine; blend in 3 tablespoons flour, 1/2 teaspoon each salt and sugar, and 1/4 teaspoon ground nutmeg; cook until bubbly. Measure the reserved cooking liquid and add milk, if necessary, to make 3/4 cup liquid; gradually add liquid to flour mixture. Cook, stirring constantly, until thickened and smooth. Blend in spinach and cook until heated through. Stir in 3/4 teaspoon lemon juice and serve. Makes 6 to 8 servings.

## Butter-Steamed Spinach Piquant

Melt 2 tablespoons butter in a wide heavy frying pan or electric frying pan. Add 2 pounds washed and slightly damp spinach; stir well. Cover and cook over medium heat until wilted. Blend in 2 teaspoons sugar and 3 tablespoons whipping cream. Simmer rapidly, stirring until liquid is almost gone. Add 1 to 2 tablespoons vinegar and salt to taste. Sprinkle with a dash ground nutmeg and serve. If desired, garnish with sliced or chopped hard-cooked egg. Makes 5 to 6 servings.

## Spinach with Cream Cheese Sauce

Melt 1 tablespoon butter or margarine in a wide, heavy frying pan or electric frying pan. Add 2 pounds washed and slightly damp spinach; stir well. Cover and cook over medium heat until barely wilted. Blend in 1 package (3 oz.) diced cream cheese, 1/2 to 3/4 teaspoon Worcestershire, and stir until cheese melts. Season to taste with salt and pepper. Makes 5 to 6 servings.

## Fresh Spinach Casserole

2½ pounds spinach
1½ teaspoons salt
2 tablespoons melted butter or margarine
2 eggs, slightly beaten
1 cup milk
⅛ teaspoon pepper
1 teaspoon finely grated onion
½ cup shredded Swiss cheese

Trim stems from spinach and wash leaves well. Put into a large pan or Dutch oven. Add 1 teaspoon salt, cover, and cook quickly in the water that clings to the leaves until tender and bright green (about 3 minutes). Drain well in a colander, pressing out water lightly with the back of a spoon; chop finely.

In a bowl, gently mix the spinach, remaining 1/2 teaspoon salt, butter, eggs, milk, pepper, onion, and cheese. Pour into a shallow, 1-quart baking dish (cover and refrigerate if done ahead). Uncover and bake in a 325° oven until set (about 30 minutes; 45 minutes if refrigerated). Serve immediately. Makes 6 servings.

## Hot Chafing Dish Spinach Salad

1 pound spinach
5 slices bacon
1½ tablespoons red wine vinegar
¼ teaspoon *each* sugar and dry mustard
Salt and freshly ground pepper

Trim, wash, and drain spinach. Tear leaves in bite-sized pieces, put in a bowl, cover, and refrigerate until time to serve. Fry bacon until crisp, drain well, and crumble into a small serving bowl; measure 2 tablespoons of the bacon drippings into a chafing dish. In a small pitcher, combine the vinegar, sugar, and dry mustard. On a tea cart or large serving tray, assemble the spinach, vinegar sauce, salt, pepper, bacon, salad servers, and chafing dish with bacon drippings.

At the table, place chafing dish over direct heat until

bacon drippings are hot. (Or for faster results, heat bacon drippings in the kitchen and bring hot to the table.) Place half of the spinach in the chafing dish and mix gently, lifting greens from the bottom of the pan to coat leaves with drippings. As spinach wilts, continue to add greens until all the spinach is in the chafing dish; keep mixing greens with drippings until leaves are evenly wilted but still bright green. Mix vinegar sauce into spinach. Season with salt and pepper; sprinkle bacon over top and serve. Makes 2 to 4 servings.

## Spinach Pie

1 small package (3 oz.), cream cheese, softened
1 cup half-and-half (light cream)
½ cup soft bread cubes, lightly packed
¼ cup shredded Parmesan cheese
2 eggs, slightly beaten
1 cup cooked spinach (about 1¼ lb. fresh), well drained and finely chopped
4 tablespoons (⅛ lb.) butter or margarine
1 large onion, finely chopped
½ pound mushrooms, finely chopped
1 teaspoon tarragon leaves
About ¾ teaspoon salt
1 unbaked 9 or 10-inch pie shell

In a mixing bowl, mash cream cheese with a fork and gradually blend in half-and-half. Add bread cubes, Parmesan cheese, and eggs to cream cheese mixture and beat with a rotary mixer or wire whip to break up bread pieces. Stir in the spinach.

Melt butter in a wide frying pan and cook onion and mushrooms until lightly browned, stirring frequently; add tarragon when vegetables are tender. Blend hot vegetables with spinach mixture. Salt to taste.

Pour vegetable filling into pastry shell. Bake on lowest rack in a 400° oven for 25 minutes or until crust is well browned. Let stand 10 minutes before serving or let cool completely and cut to serve. Makes 6 to 8 servings.

## Joe's Special

2 pounds ground beef chuck, crumbled
2 tablespoons olive oil or salad oil
2 medium-sized onions, finely chopped
2 cloves garlic, minced or mashed
½ pound mushrooms, sliced (optional)
1¼ teaspoons salt
¼ teaspoon *each* ground nutmeg, pepper, and oregano
1 package (10 oz.) frozen chopped spinach thawed and well drained, or ½ pound fresh spinach, washed, drained and chopped (about 4 cups)
4 to 6 eggs

Brown ground beef well in oil in a large frying pan over high heat. Add onions, garlic, and mushrooms, if desired; reduce heat and continue cooking, stirring occasionally, until onion is soft. Stir in seasonings and spinach; cook for about 5 minutes longer. Add eggs; stir mixture over low heat just until eggs begin to set. Makes 4 to 6 servings.

## Spinach Soufflé with Chive Sauce

2 packages (10 oz. *each*) frozen chopped spinach, thawed
4 eggs
1 cup whipping cream
½ cup melted butter or margarine
½ teaspoon salt
¼ teaspoon ground nutmeg
About ¼ cup shredded Parmesan cheese
Chive sauce (recipe follows)

In a blender, whirl spinach with eggs until very smooth, then add cream, butter, salt, and nutmeg.

Heavily butter a 1-quart soufflé dish (or a straight-sided baking dish) and dust with 2 tablespoons of the cheese. Pour in spinach mixture; dust with remaining cheese. Cover the dish and chill as long as overnight. Bake in a 350° oven for 1 hour 5 minutes or until the center feels firm when touched. Serve at once, passing the cold chive sauce to spoon over each portion. Makes 6 to 8 servings.

**Chive Sauce.** In a mixing bowl, gradually blend 3/4 cup whipping cream with 1 package (3 oz.) cream cheese, then whip stiff. Stir in 2 tablespoons chopped fresh or freeze-dried chives. Chill, covered.

## Stir-Fry Spinach, Tomato, and Bacon

1 pound fresh spinach
3 slices bacon
1 small onion, finely chopped
1 clove garlic, minced
2 tablespoons regular strength chicken broth or water
1 basket cherry tomatoes, halved
½ teaspoon salt
Dash pepper

Wash spinach thoroughly, discarding stems; stack leaves and cut crosswise into 1/2-inch-wide strips. Set aside.

Cook bacon in wok or large frying pan over low heat until crisp; remove and drain. Pour out bacon drippings and save. Don't wash pan. Before starting to cook, prepare all remaining ingredients and place within reach of your range. Place pan over high heat and put in 2 tablespoons of the bacon drippings. When hot, put in onion and garlic. Stir with a wide spatula and fry for about 1 minute. Add spinach and broth; stir to mix with onion, cover pan, and cook 1/2 to 1 minute until spinach is just wilted. Add tomatoes, salt, and pepper; stir and fry until tomatoes are just hot (less than 1 minute). Spoon into a warm serving dish and sprinkle with crumbled bacon. Makes about 6 servings.

# TURNIP TOPS

The green tops from tender, young turnips make a tasty green vegetable. Remove heavy part of stems.

### Butter-Steamed Turnip Tops with Bacon

In a wide heavy frying pan or electric frying pan, cook 4 slices diced bacon until browned. Drain off all but 2 tablespoons drippings; stir in 4 cups chopped turnip tops and 2 tablespoons regular strength beef broth. Cover and cook over medium heat, stirring occasionally, for about 10 minutes. If desired, add 2 tablespoons shredded sharp Cheddar cheese and stir until melted. Makes 4 servings.

### Turnip Top Soup

In a heavy frying pan or electric frying pan, melt 2 tablespoons butter or margarine; stir in 1 1/2 cups regular strength beef broth; bring to a boil. Add 4 cups chopped turnip tops, stirring well. Cover and simmer for about 5 minutes. Blend in 1 cup half-and-half (light cream); whirl in a blender until smooth. Return to pan and heat to simmering. Makes 4 to 6 servings.

# WATERCRESS

The sharp biting flavor of watercress leaves adds lively, spicy flavor to a variety of foods; tucked into sandwiches, in soups, combined with meats, as well as in salads.

### Cress and Beef Sandwich Spread

Place 1 jar (2 1/2 oz.) dried beef, chopped, in a wire strainer, and pour boiling water through it; drain well. Beat 2 tablespoons mayonnaise, 1 tablespoon prepared horseradish, and 4 ounces (half an 8-oz. package) soft cream cheese until fluffy. Fold in dried beef and 1/2 cup finely chopped watercress. Makes about 1 cup spread.

## Watercress-Cheese Sandwiches

Combine 1/2 cup finely minced watercress, 1 small package (3 oz.) cream cheese, 1/2 cup chopped walnuts, and 2 hard-cooked eggs, chopped very fine. Mix well and add salt to taste. Spread between buttered slices of bread. Makes six sandwiches.

## Watercress Biscuits

In a bowl, combine well 3 cups prepared biscuit mix, 1/2 cup finely minced watercress, and 1 tablespoon grated onion. Stir in 3/4 cup milk to make a stiff dough. Turn onto a floured board and knead 4 or 5 times. Roll out to 1/2-inch thickness. With a long straight-edged knife, cut into 1-inch diamonds. Place on ungreased baking sheet; brush with melted butter. Bake in a 375° oven for 12 to 15 minutes. Serve hot. Makes 60 biscuits or enough for 30 servings of 2 biscuits.

# JERUSALEM ARTICHOKE

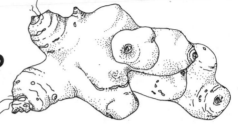

## Selection and Storage

**Selection:** Jerusalem artichoke is a potato-like tuber that grows underground. It resembles ginger root in appearance only. Choose firm tubers that are free from mold.

**Buying:** Allow 1 to 1 1/2 pounds for 4 servings.

**Storage:** Place in a plastic bag and refrigerate until ready to use. They will keep a week to 10 days.

## Cutting and Cooking

To prepare artichokes, scrub each tuber well with a very stiff brush or peel with vegetable peeler. You may have to break tubers apart to reach the skin between the protuberances. Any bits of skin that remain will resemble, when cooked, the skin of new potatoes. If recipes suggest peeling, keep artichokes from discoloring by covering them with cold water. The tubers can be sliced or diced before cooking or cooked whole and then cut.

**To cook Jerusalem artichokes,** drop the whole or cut pieces into a small amount of boiling salted water. Cover and cook until tender (15 to 30 minutes for whole tubers, 8 to 12 minutes for sliced or diced tubers); drain.

## Seasoning and Serving

Jerusalem artichokes can be served either raw or cooked. They have a sweet, nut-like taste when raw.

**To serve raw,** peel the tubers, cut into sticks, and crisp in ice water. Sprinkle them with salt as you would radishes or serve them with a savory dip (see recipes for raw vegetable dips page 7 ). When pieces of raw (or partly cooked) Jerusalem artichokes are mixed with other foods, they contribute a flavor and texture similar to water chestnuts.

**Cooked,** they are delicious when served with salt, pepper, butter, and a touch of lemon juice.

## Butter-Steamed Jerusalem Artichokes

Cut 1 pound scrubbed artichokes into slanting slices about 1/8 inch thick (you should have about 3 cups). Melt 2 tablespoons butter or margarine in a wide frying pan or electric frying pan over moderate heat. Add sliced artichokes, 2/3 cup water, 1/4 teaspoon fines herbes (or crushed tarragon leaves), and 1 teaspoon lemon juice.

Stir mixture together, cover, and cook over moderate heat for 10 to 12 minutes or until artichokes are just tender and liquid is absorbed. Season to taste with salt and pepper; sprinkle with chopped parsley or minced chives. Serve immediately. Makes 3 to 4 servings.

## Broth-Steamed Jerusalem Artichokes

Follow directions above for Butter-Steamed Jerusalem Artichokes using regular strength beef broth instead of water.

## Scalloped Jerusalem Artichokes

Scrub well, or peel 2 pounds Jerusalem artichokes; slice thinly (you should have about 5 cups). Rub a cut clove garlic over inside of a 1 1/2-quart casserole, then butter it lightly. Mix 2 tablespoons flour, 1/2 teaspoon salt, 1/4 teaspoon paprika, and a dash white pepper. Layer artichokes in casserole, using about 1/4 at a time; sprinkle each of the

3 lower layers evenly with about 1/3 of the flour mixture.

Mix 1 teaspoon lemon juice with 1/4 cup whipping cream; pour over artichokes. Cover and bake in a 325° oven for about 1 hour and 15 minutes or until artichokes are tender; gently stir once after about 45 minutes.

Uncover, sprinkle with 1/2 cup shredded Gruyère, Samsoe, or Swiss cheese, place under broiler, and broil for 3 to 5 minutes until cheese is bubbly and golden. Serve sprinkled with chopped parsley. Makes 6 servings.

## Jerusalem Artichoke Provençal

1½ pounds Jerusalem artichokes, scrubbed
    Boiling salted water
2 tablespoons olive oil or salad oil
1 clove garlic, crushed
1 pound (about 4) tomatoes, peeled, seeded, and diced
    Salt and pepper to taste
    Minced parsley

Into a saucepan of rapidly boiling water, drop whole artichokes; boil for only 5 minutes; drain and dice. In a wide saucepan heat oil with garlic; add tomatoes and simmer for 10 minutes. Discard garlic and add artichokes; simmer until tender (about 10 minutes more). Season with salt and pepper, sprinkle with minced parsley, and serve. Makes 4 to 6 servings.

## Pan-Fried Jerusalem Artichokes

1½ pounds Jerusalem artichokes, scrubbed and sliced or diced thinly
4 to 6 tablespoons butter or margarine
    Salt to taste
    Minced parsley

Heat 4 tablespoons butter in a frying pan over medium-high heat. Put in artichokes and sauté quickly, stirring, until lightly browned and tender (about 8 to 10 minutes). Season to taste with salt, pepper, and sprinkle with minced parsley. Makes 4 to 6 servings.

## French Fried Jerusalem Artichokes

1½ pounds Jerusalem artichokes
    Salt and pepper
    All-purpose flour
1 egg, beaten
    Fine cracker crumbs
    Salad oil for deep frying

Scrub Jerusalem artichokes and cut in thin slices with as large a surface as possible. Sprinkle lightly with salt and pepper, coat with flour, and shake off excess. Dip in beaten egg, drain briefly, then roll in cracker crumbs. Place on a rack and dry for about 30 minutes or longer. In a deep pan or deep fat fryer, heat 1 inch or more of the oil to 365°; drop a few artichoke slices in at a time and cook until well browned on both sides. Remove with a slotted spoon, drain, sprinkle with additional salt, if needed. Serve hot. Makes about 6 servings.

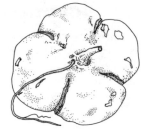

# JICAMA

### Selection and Storage

**Selection:** Jicama (*hee*-cah-muh) is a brown root shaped something like a rutabaga but much larger; those in markets weigh 1 to 6 pounds. Choose well formed jicama free from blemishes.

**Buying:** Allow 1 pound for 4 to 6 servings.

**Storage:** Store, unwashed, in the refrigerator. Will keep for several weeks. Tightly cover cut pieces with clear plastic film.

### Cutting and Cooking

Peeled jicama is white, juicy, mild, and crisp. It is delicious served raw as an appetizer or in salads, and it combines nicely with either fruits or other vegetables. Cooked, jicama retains its crisp texture when added to braised or sautéed dishes. It can also be substituted for water chestnuts.

Wash and peel jicama, then thinly slice, cut into squares, or dice it. Cook or serve raw according to individual recipe directions.

### Jicama Appetizer

Blend 1 tablespoon salt with 1/4 teaspoon chile powder and put in a small bowl. Also cut 1 lime in wedges. Peel and slice a 1 to 2-pound piece of jicama, cut in 1/4 to 1/2-inch-thick slices, and arrange on a serving tray. To eat, rub lime over surfaces, then dip in salt. Serves 6 to 8.

### Jicama with Fried Potatoes

Peel 2 large (about 1 lb.) baking potatoes and cut in 1/4-inch thick slices, about 1 1/2 inches square (you should have about 3 cups).

In a wide frying pan over medium heat, melt 6 tablespoons butter or margarine. Add potatoes, onions, and jicama. Cook, turning as needed with a wide spatula, until potatoes are tender and golden (take 25 to 30 minutes). Stir in 1/2 teaspoon salt, 1/8 teaspoon pepper, and 1/2 cup minced green onion, including part of tops, and serve. Makes about 4 servings.

### Jicama and Fruit Compote

    1 pound jicama
    4 lemon slices
    ½ cup sugar
  2½ cups water
    3 or 4-inch stick of cinnamon
    2 apples, peeled and sliced
  1½ cups (about 6 oz.) dried apricots

Wash, peel, and thinly slice jicama then cut in bite-sized squares. In a pan, combine the jicama, lemon slices, sugar, water, and cinnamon stick. Bring to a boil; simmer, uncovered, 5 minutes. Add apples to pan with dried apricots. Simmer 2 to 4 minutes longer until apples are tender. Remove from heat; chill. Discard lemon slices. Makes 4 to 6 servings.

### Pico de Gallo

  2 cups peeled, diced jicama
  1 green pepper, seeded and slivered
  ½ medium-sized mild onion, thinly sliced
  1 cup sliced or diced cucumber
  ¼ cup olive oil
  2 tablespoons white or red wine vinegar
  ½ teaspoon oregano leaves
    Salt and pepper to taste

Combine jicama, green pepper, onion, and cucumber in a bowl. Mix olive oil, vinegar, and oregano and pour over vegetables. Mix lightly. Salt and pepper to taste. *Pico de gallo* (rooster's beak) makes a crunchy, savory salad. Makes 4 to 6 servings.

 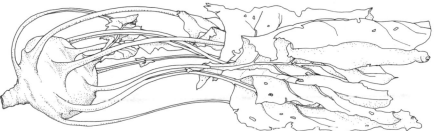

# KOHLRABI

### Selection and Storage

**Selection:** Choose small to medium-sized bulbs with fresh green leaves. Avoid deeply scarred or blemished kohlrabi.

**Buying:** Allow 1 medium bulb for each serving.

**Storage:** Keep enclosed in a plastic bag, refrigerated, until ready to use. Keeps several days.

### Cutting and Cooking

Both the bulb-shaped section of the stem and the leaves that shoot out of it are good to eat. If kohlrabi is very young and tender, it is not necessary to peel it before cooking. Peel more mature kohlrabi deep enough to remove tough outer skin. Cut off the leaves and stems, dice stems (leaves can be cooked as in recipe below).

**To cook kohlrabi,** drop cut pieces into a small amount of boiling salted water. Cut kohlrabi bulb into cubes, slices, or sticks. Cook until tender (about 12 to 25 minutes); drain.

### Seasoning and Serving

The crisp white flesh of the bulb is delicious either raw or cooked. To serve raw, cut in slices or sticks to munch, include as an appetizer to dip in mayonnaise or sour cream mixtures, or serve on relish trays.

Cooked, it can be seasoned simply with salt, pepper, butter or margarine, or a sprinkle of vinegar or lemon juice.

### Boiled Kohlrabi with Leaves

Cut leaves and stems from 1 1/2 pounds kohlrabi. Remove and discard stems; chop the leaves and reserve. Peel the bulb section, if necessary; dice and cook, covered, in 1 cup salted regular strength chicken broth for 12 to 20 minutes or until tender. In another pan, cook chopped leaves, covered, in 1/2 cup salted water until wilted (about 3 minutes). Drain leaves and combine with cooked kohlrabi; dress with 2 tablespoons lightly browned butter. Season to taste with salt and pepper. Makes 4 to 6 servings.

### Butter-Steamed Kohlrabi

In a heavy frying pan or electric frying pan melt 3 tablespoons butter or margarine over high heat. Put in 3 cups very thinly sliced, peeled kohlrabi (cut large slices in halves or quarters), 7 tablespoons water, and 1/2 teaspoon basil leaves. Cover and cook quickly, removing cover and stirring frequently. Cook about 8 minutes, or until vegetable is tender. Sprinkle with 1 tablespoon minced chives or sliced green onion tops, and salt and pepper to taste. Makes 4 to 6 servings.

---

## Kohlrabi La Ronda

    About 2½ pounds kohlrabi
    Boiling salted water
  4 tablespoons butter or margarine
1½ tablespoons all-purpose flour
1½ cups milk
  1 cup shredded Cheddar cheese
    Salt and pepper
  1 tablespoon chopped parsley
    Paprika

Cut off stems of kohlrabi and with a sharp knife peel off tough fibrous skin. Cut into 1/8-inch slices. Cook in boiling salted water to cover until just tender (5 to 10 minutes); drain and turn into a 1 1/2-quart baking dish.

In a saucepan melt the butter over medium heat; stir

in flour and cook until bubbly. Remove from heat and gradually add milk, stirring constantly. Cook, stirring, until thickened, then remove from heat and stir in cheese until melted. Season with salt and pepper to taste. Pour the sauce over the kohlrabi and sprinkle on the parsley and paprika. Bake, uncovered, in a 350° oven until sauce bubbles (about 20 minutes). Serve immediately. Makes about 6 servings.

## Kohlrabi Salad Niçoise

2 cups coarsely shredded peeled kohlrabi
¼ cup sliced radishes
3 tablespoons mayonnaise
1 teaspoon sugar
2 tablespoons *each* vinegar and salad oil
½ teaspoon dry mustard
  Salt and pepper to taste
  Lettuce cups
  Paprika

In a bowl, combine kohlrabi and radishes. Combine the mayonnaise, sugar, vinegar, salad oil, and mustard; add to salad and mix lightly. Season with salt and pepper. Chill several hours. Serve in crisp lettuce cups sprinkled with a little paprika. Makes 3 to 4 servings.

## Selection and Storage

**Selection:** Different varieties may be white, brown, or cream colored. When caps are closed so that the gills underneath are not visible, it indicates mushrooms are newly picked. Moisture evaporates during storage and caps gradually open so the brown gills can be seen. When appearance is important—such as when serving mushrooms raw—closed mushrooms are preferred; open ones are fine for cooking and may be more economical purchased by the pound as these mushrooms weigh less. Size has nothing to do with maturity or flavor, but for some uses uniform mushrooms of a particular size are required. Avoid mushrooms that are wilted and slick or dry and wrinkled.

**Buying:** Allow 1 pound for 4 to 6 servings.

**Storage:** Store, unwashed, in a plastic bag (unsealed to allow air circulation) in the refrigerator. They will keep several days.

## Cutting and Cooking

When ready to use, rinse briefly in cool water and pat dry, or wipe mushrooms with a damp paper

towel—never soak them. If the end of the stem is gritty, slice it off.

**For sliced mushrooms,** cut lengthwise through caps and stems into thin or thick slices.

**Small mushrooms** may be left whole, or cut lengthwise through stems into halves or quarters.

**Large mushrooms,** with stems removed, may be stuffed for appetizers or entrées.

Simple sautéing in butter over medium heat takes just a few minutes and is the basic technique used in many of the following recipes.

## Seasoning and Serving

Mushrooms are delicious served raw as appetizers, in salads, or as a vegetable serving.

Cooked, mushrooms can be used in an infinite number of ways. They can be added to meats, poultry, other vegetable dishes, and soups. Here are some recipes for times when they can be used alone.

### Sautéed Mushrooms

Clean mushrooms as directed above and slice through cap and stem, quarter, or leave whole if small. Melt 3 to 4 tablespoons butter for every 1/2 pound mushrooms in a heavy frying pan or electric frying pan. Sauté until tender (about 5 to 8 minutes). Season with salt, pepper, and minced parsley or tarragon leaves. They can be served on toast or alone.

### Creamed Mushrooms

Sauté mushrooms as directed on page 60 using 2 tablespoons butter for each 1/2 pound mushrooms. When tender, stir in 1 cup Béchamel Sauce (recipe page 5) or sour cream for *each* 1/2 pound mushrooms. Season to taste with salt and pepper, and sprinkle with minced parsley, if desired.

### Broiled Mushrooms

Pick out uniform medium-sized (about 1 1/2 inches diameter) mushrooms. Clean them as directed under *cutting and cooking*, page 60, and remove stems. (Use stems in recipes calling for chopped mushrooms.) Brush mushrooms all over with butter, olive oil, or a mixture of half melted butter and half olive oil. Sprinkle lightly with salt and pepper. Broil about 4 inches from heat, turning and basting as needed until tender and browned (about 5 to 8 minutes).

### Mushroom Butter

Sauté 1/2 pound finely chopped mushrooms in 3 tablespoons butter for 8 minutes; add 1 tablespoon Sherry, 1/4 teaspoon salt, dash pepper, and 1/2 cup (1/4 lb.) soft butter. Cream together until well blended. Add about 1/2 garlic clove or a pinch minced tarragon leaves, if desired.

Serve on steaks, broiled fish, or as a spread for turkey breast sandwiches.

### Mushrooms Paprika

Clean 1 pound mushrooms. Slice or quarter them if they are large. Put in a heavy saucepan with 1/4 cup (1/8 lb.) butter and 1 cup chopped or sliced onions. Cook slowly until onions are soft (about 5 minutes). Blend in 2 tablespoons all-purpose flour, 1 tablespoon paprika, and a dash liquid hot pepper seasoning. Cook, stirring, for 4 minutes. Add 1 cup sour cream and heat (do not boil). Serve on rice or baked potatoes. Makes enough for about 6 servings.

### Pickled Mushrooms

Cook 3 pounds cleaned whole or quartered mushrooms in 1 quart water for 15 minutes. Drain, reserving the liquid. Slice 2 medium-sized onions very thin and arrange mushrooms and onions in layers in a crock or jars. Simmer 1/2 cup distilled white vinegar, the mushroom liquid, 1 1/2 teaspoons salt, 1/2 bay leaf, and 1/4 teaspoon whole pepper for 10 minutes, then strain over mushrooms. Float 1 teaspoon olive oil over top and chill for at least 24 hours before serving.

## Raw Mushroom Dip

    1 package (3 oz.) cream cheese, softened
    ½ cup sour cream
    ½ teaspoon salt
    ¼ teaspoon tarragon
    1 clove garlic, mashed
    1 tablespoon minced green onion, including part of tops
    1½ cups finely chopped mushrooms (about ¼ lb.)
        Crisp lettuce
        Chopped parsley

In a small bowl, mix together cream cheese, sour cream, salt, tarragon, garlic, and onion. Add mushrooms and stir until well blended. Chill at least 2 hours before serving. To serve, spoon the dip into a bowl lined with crisp lettuce; garnish with chopped parsley. Offer corn chips or any fairly sturdy chips or crackers for dipping. Makes about 2 1/2 cups dip.

## Marinated Mushroom Platter

    1½ pounds mushrooms
    ½ pound Swiss or Gruyère cheese
    2 cups thinly sliced celery
    ⅓ cup lemon juice
    ½ cup olive oil or salad oil
    ¾ teaspoon tarragon leaves, crumbled
    ¼ teaspoon *each* dry mustard, salt, and garlic salt
    ⅛ teaspoon pepper
        Parsley

Clean mushrooms; pat dry. Cut mushrooms through stems into thin slices; place in a bowl. Cut cheese into matchstick-sized slivers; place in another bowl with celery.

In a small jar or bowl, combine the lemon juice, oil, tarragon, mustard, salt, garlic salt, and pepper. Shake or stir to blend. Pour half the dressing over the mushrooms and the remainder over the cheese and celery; mix. Cover each and refrigerate overnight, stirring gently several times.

Using a slotted spoon, lift mushrooms, cheese, and celery from marinade and arrange on a shallow platter. Garnish with parsley. Drizzle with remaining marinade. Makes about 10 servings.

## Mushrooms au Gratin

    1 pound mushrooms
    2 tablespoons butter or margarine
    ⅓ cup sour cream
    ¼ teaspoon salt
        Dash pepper
    1 tablespoon all-purpose flour
    ¼ cup finely chopped parsley
    ½ cup shredded Swiss or mild Cheddar cheese (about 2 oz.)

Clean mushrooms and slice lengthwise through the stems into about 1/4-inch-thick slices. In a large frying pan, heat the butter over medium-high heat. Sauté mushrooms, stirring until lightly browned. Cover pan for about 2 minutes until they start losing juices.

Blend the sour cream with salt, pepper, and flour until smooth. Stir into the mushrooms and heat, stirring, until blended and mixture begins to boil. Remove from heat and pour into a shallow, rimmed, heatproof dish. Sprinkle parsley and cheese evenly over top. This much can be done ahead and refrigerated.

Shortly before serving, place, uncovered, in a 425° oven until mushrooms are heated through and the cheese is melted (about 10 minutes). Serves 4.

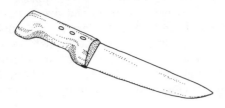

# Fresh Mushroom Salad

½ pound mushrooms, thinly sliced through the stems
⅔ cup salad oil or olive oil
⅓ cup wine vinegar
¼ teaspoon tarragon leaves
⅛ teaspoon ground nutmeg
¾ teaspoon salt
    About 3 cups butter lettuce, in bite-size pieces
    About 2 cups spinach leaves, stems removed, in bite-size pieces

Place mushrooms in a deep bowl. Pour the salad oil, wine vinegar, tarragon, nutmeg, and salt over the mushrooms. Mix gently, cover, and let stand at room temperature for about an hour, stirring occasionally.

Combine lettuce and spinach with mushrooms and dressing and mix lightly to serve. Makes 5 servings.

# Swedish Mushroom Sauce

You might serve this versatile sauce over crisp toast, with omelets, meats, or poultry.

½ pound mushrooms, sliced
3 tablespoons *each* butter or margarine and flour
1 teaspoon salt
¼ teaspoon white pepper
1½ cups half-and-half (light cream)
1½ tablespoons *each* lemon juice and dry Sherry

In a frying pan over medium heat, sauté mushrooms in butter for about 5 minutes. Stir in flour, salt, and pepper. Add half-and-half; cook, stirring constantly, until thick and smooth. Stir in lemon juice and Sherry. Makes about 2 1/2 cups sauce.

# Scalloped Mushrooms

1½ pounds fresh mushrooms, washed, drained, and sliced
3 cups soft French bread crumbs (whirl fresh bread pieces in blender)
¾ cup butter or margarine, melted
    Salt and pepper
½ cup dry white wine

Place about a third of the mushrooms in a buttered 2-quart baking dish; cover with about a third of the bread crumbs and drizzle about a third of the butter over the crumbs. Sprinkle with salt and pepper. For the top layer, cover with remaining mushrooms, sprinkle them with salt and pepper, and pour wine over all.

Cover and bake in a 325° oven for about 35 minutes. Mix remaining butter and crumbs and spoon over mushrooms. Bake, uncovered, 10 more minutes or until crumbs are toasted. Makes about 12 servings.

# Much-More-Than Mushrooms

1 pound mushrooms, sliced
4 tablespoons (⅛ lb.) butter or margarine
½ cup *each* chopped green onions (including part of tops), chopped celery, and chopped green pepper
¾ teaspoon salt
¼ teaspoon pepper
2 tablespoons chopped parsley
½ cup mayonnaise
6 slices firm white bread
3 eggs
2 cups milk
¼ cup grated or shredded Parmesan or Romano cheese

In a frying pan over medium-high heat, sauté mushrooms in butter for 5 minutes. Add the green onions, celery, green pepper, salt, pepper, and parsley; cook about 3 minutes.

Remove from heat, stir in the mayonnaise, and set aside.

Remove crusts from bread and cut into 1-inch squares. Put half of the bread in a greased 2 1/2-quart casserole. Spoon mushroom mixture over bread, then cover with remaining bread. Beat eggs and milk together until frothy; pour over mixture in casserole.

Refrigerate, covered, for at least 1 hour or as long as overnight.

Bake, uncovered, in a 325° oven for 50 minutes. Remove from oven and sprinkle cheese over top; return to oven and bake 10 minutes longer or until golden brown. Makes 6 servings.

# Mushroom Bouillon

2 pounds finely chopped mushrooms
2 tablespoons chopped onion
2 tablespoons melted butter
2 quarts regular strength chicken broth
4 tablespoons Sherry (optional)
    Salt
    A few drops bottled brown gravy sauce

In a large saucepan cook mushrooms and onion in melted butter, stirring, until vegetables look moist but are not browned. Add chicken broth, cover, and simmer for 30 minutes.

Moisten and wring dry a muslin cloth; line a wire strainer with it and set over a deep bowl. Pour mushroom mixture through cloth; when cool to touch, squeeze out all juices possible. Discard mushrooms; season broth with Sherry and salt, if desired, and add a few drops gravy sauce to enrich color. Serve in cups to sip. Makes 2 quarts or 10 to 12 servings of about 3/4-cup size.

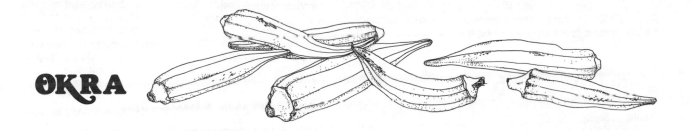

# OKRA

## Selection and Storage

**Selection:** Choose okra (sometimes called gumbo) with crisp, bright green pods, free from blemishes. Select small to medium-sized pods.

**Buying:** Allow 1/4 pound for each serving.

**Storage:** Keeps best at about 50°. Will keep for several days in crisper of refrigerator.

## Cutting and Cooking

Wash and trim off stem ends of okra. Leave small pods whole. Cut large pods into 1/2-inch slices. It is important to cook okra just until tender so that it retains bright green color and slight crispness. When overcooked, okra becomes slippery and has a dull green color.

**To cook okra,** drop into enough boiling salted water to cover. Boil just until tender (5 to 10 minutes); drain immediately.

## Seasoning and Serving

For these two simple ways to serve okra below, you can use fresh okra, cooked following directions under *cutting and cooking* at left, or use frozen okra, thawed and cooked following the directions on the package.

### Buttered Okra

Trim and cook okra following directions under *cutting and cooking* at left; drain. Season to taste with melted butter, salt, pepper, and a sprinkling of lemon juice. Serve hot.

### Okra with Herbs

Prepare okra same as above for Buttered Okra, adding for each cup of vegetable 1 tablespoon *each* minced parsley and chives and 1/4 teaspoon thyme leaves. Mix well and serve hot.

# Bhindi Masala (Stuffed Okra)

    2 packages (10 oz. *each*) frozen okra
    2 tablespoons ground coriander
    1 teaspoon salt
    1½ teaspoons ground turmeric
    ½ teaspoon *each* pepper and ground cumin
      About 3 tablespoons *each* butter and shortening or
        salad oil
    1 can (1 lb.) small white onions, drained, (optional)

Allow the okra to thaw at room temperature; wipe each dry. With the tip of a small sharp knife, cut a slit in the side of each okra. Combine the coriander, salt, turmeric, pepper, and cumin; spoon about 1/4 teaspoon (or a little less) of the spice mixture into each okra (you should have about one quarter of the mixture left). This much of the preparation can be done as much as 3 hours ahead; keep the okra refrigerated until cooking time.

Just before serving, heat about 2 tablespoons *each* butter and shortening in a frying pan over medium-high heat; add the okra and stir-fry about 7 to 8 minutes or until tender, adding more butter and shortening as needed. Spoon okra into a warm serving dish.

If onions are used, stir the remaining spice mixture into the butter remaining in the pan, adding the drained onions, and stirring until lightly browned and heated through. Add to the okra in the serving dish. Makes about 6 to 8 servings.

# Stewed Okra with Onions

Fresh coriander lends a distinctive flavor to this Lebanese dish. If fresh coriander (also called cilantro and Chinese parsley) isn't available, use 1/2 cup finely chopped parsley and 1/2 teaspoon whole coriander seed.

    1 package (10 oz.) frozen okra or ½ to ⅔ pound fresh okra
      Boiling water
    6 to 8 small boiling onions, peeled
    2 tablespoons olive oil
    4 tablespoons water
    ½ cup finely chopped fresh coriander
    1 small clove garlic, minced
    2 tablespoons lemon juice

Let the okra thaw; drop into a saucepan of boiling water to cover and cook 2 to 3 minutes or just until okra is easy to pierce. If using fresh okra, trim stem ends; cook in boiling water to cover for 5 minutes. Drain okra at once and immerse in ice water to cool quickly. When cold, drain and set aside.

Place onions in a medium-sized frying pan with olive oil and water. Cover and cook over moderately-low heat until the water evaporates, then continue cooking, turning onions occasionally until they are lightly browned (about 10 minutes).

Uncover pan and add fresh coriander (or parsley and coriander seed) and garlic. Add the okra and cook, stirring, until okra is warm. Place in a serving dish and serve hot or at room temperature. Just before serving, drizzle lemon juice over the vegetables. Makes 6 to 8 servings.

## Erik's Okra Soup

6 cups regular strength chicken broth (or 6 cups water and 4 chicken bouillon cubes)
1 cup diced cooked ham (bone optional)
½ cup diced cooked chicken (optional)
¼ teaspoon thyme leaves
1 bay leaf
1 small, dried hot chile pepper
½ cup *each* sliced celery and diced onion
1 medium-sized tomato, diced
1 package (10 oz.) frozen okra, partially thawed
3 tablespoons salad oil
2 tablespoons all-purpose flour
1 can (4½ oz.) shrimp, rinsed and drained
    Steamed rice
    Gumbo filé powder
    Liquid hot pepper seasoning (optional)

Combine broth, ham, chicken (if used), thyme, bay leaf, chile pepper, celery, onion, and tomato in a pan that can be covered. Bring to a boil, reduce heat, cover, and simmer about 10 minutes. Cut okra into 1/2-inch thick pieces and sauté in salad oil over medium heat 5 minutes. Sprinkle flour over okra, stirring to mix in, and simmer 5 minutes; set aside. About 5 minutes before

serving, remove chile and bay. Add okra and shrimp; cook just until heated through and okra is tender.

Partially fill individual soup bowls with hot steamed rice. Ladle gumbo over rice and sprinkle gumbo filé powder on top to taste. Add a dash of liquid hot pepper seasoning, if desired. Makes 6 to 8 servings.

## Okra with Tomatoes

1 pound okra or 1 package (10 oz.) frozen okra
2 tablespoons butter or margarine
1 medium-sized onion, chopped
3 medium-sized tomatoes, peeled, seeded, and diced (or 1½ cups canned tomatoes)
½ teaspoon salt
1 small dried hot chile pepper, crushed, or 5 drops liquid hot pepper seasoning
    Dash pepper

Remove stems from the fresh okra and wash well; cut in 1/2-inch slices (cut the frozen okra while still frozen). Heat butter in a large saucepan or in a frying pan with a tight fitting lid. Add the okra and onion to the pan and cook over medium heat, stirring frequently until the onion starts to brown (okra will not brown). Remove from heat and add the tomato, salt, chile pepper, and pepper.

Return to heat, bring to boiling, then reduce heat, cover, and simmer slowly until the okra is tender (5 to 10 minutes). Serve immediately. Makes about 4 generous servings.

# ONIONS & LEEKS

## Selection and Storage

**Selection:** Choose *dry onions* that look fresh, firm, and have well-shaped bulbs. They should have slim necks and the dry skin should crackle. Avoid dark-spotted onions. Firm onions with a brown stain on the outer skins are not necessarily injured; the stain will probably peel away.

These general guidelines can help predict the nature of a dry onion. Flat, early spring onions usually hold more moisture, are slightly springy, have few outer skins, and are sweet and mild. Later in the season, the more globular onions are firmer and their outer skins are thinner and dryer; these onions store well and are more pungent in flavor. Two major exceptions to this rule are the big spherical Spanish-type onions (sweet and mild, available in mid-winter) and the most vertical onion—the elongated Italian red onion (moist and mild, a spring onion). Choose chives, green onions, and leeks that have crisp green tops.

**Buying:** Allow 1/4 pound dry onions for each serving or 1 sweet onion per serving. A pound of leeks serves 2 to 3.

**Storage:** Store dry onion varieties in a cool, dry, dark airy place to prevent sprouting and decay.

Green onions and leeks should be rinsed and dried; enclose in a plastic bag and store in the refrigerator. Use within a few days.

## Cutting and Cooking Onions

Remove only the paper-like skin when you peel onions. *Slice large onions* through from top to bottom, then peel. Put the cut side down on a board, then use a French knife to slice or chop. To chop onions with fewest tears, keep cut surface against board, slice lengthwise in 1/8-inch thick slices almost to base, then slice crosswise, and finally chop ends. *Cut whole onion rings or quarters* from unpeeled onions, then remove dry skin from each.

The easiest way to remove skins from small boiling onions is to pour boiling water over them, let stand 2 or 3 minutes, then drain and peel. Scoring root end of each onion with a small cross helps keep it intact while cooking.

**To cook small, whole boiling onions,** drop the peeled and scored onions (see above) in boiling salted water to cover. Cook, uncovered, until tender (15 to 20 minutes); drain and season.

**To cook green onions,** rinse well. Remove any loose layers of skin. Trim roots and tops, leaving 2 to 3 inches of green. Cook whole or sliced lengthwise in about 1 inch boiling salted water, covered, until tender (8 to 10 minutes); drain and season.

## Onions as Seasoning

Onions have a remarkable flavor affinity for almost all meats, vegetables, fish, poultry, eggs, cheese, and even some fruits. The flavoring substance in onions is a volatile oil; it evaporates rapidly when exposed to air, so cut onions just before using them as a seasoning.

**Rubbing the sides** of a salad bowl, soufflé dish, or other baking dish with a cut onion before using, captures subtle onion flavor in the dish.

**For onion juice,** cut a crosswise slice from an onion, then with edge of spoon or knife, scrape onion from center and catch juice in a spoon.

**Sautéing chopped onion** for a few minutes in small amount of shortening or butter before adding them to other ingredients develops a rich, full flavor quite unlike raw onion. When you cook onions slowly for a long time, their raw harshness is completely converted to a mellow, rich sweetness (see recipe for Slow Cooked Onions, following).

# Slow-Cooked Onions

You can prepare these onions ahead to use for seasoning other vegetables or to top a hot sandwich such as a cheeseburger.

3 tablespoons butter or margarine
4 large onions, sliced and separated into rings

Melt butter in a 10 to 12-inch-diameter frying pan; add onions. Cook over moderate heat for about 30 minutes, stirring occasionally at first and more frequently as rings begin to develop a golden color. The onions should not show signs of browning for at least 15 minutes; if they do, reduce heat.

When onions are generally a light gold and a few bits are browned, remove from heat. Serve onions hot or chill, covered, up to 3 or 4 days, reheating as needed. Makes about 1 1/2 cups.

# Stuffed Sweet Onions

4 mild sweet onions, about 3½ inches in diameter
3 teaspoons Worcestershire
  About ½ teaspoon salt
½ cup shredded Longhorn Cheddar cheese (about 2 oz.)
4 tablespoons grated Parmesan or Romano cheese
2 tablespoons butter or margarine
  Paprika

Remove the onion skin and cut off stem and blossom end of each. With a small, sharp knife, cut a 1-inch-wide hole to center of each onion. Pour 3/4 teaspoon Worcestershire inside each onion. Sprinkle salt (about 1/8 teaspoon for *each*) into cavity and on outside of onions. Mix the cheeses. Into the center of each onion, poke about 3 tablespoons of cheese mixture, then top with 1/2 teaspoon butter. Dust paprika generously over tops of onions, then wrap each securely in heavy foil. (This much can be done ahead.)

When ready to use, set onions slightly apart on rack in the oven and cook at 425° until they feel soft when pierced with a fork (about 45 minutes). Or if you are using the barbecue, tuck onions in around the edges of low glowing coals; cooking time will be about the same. Makes 4 servings.

# Almond Onion Casserole

2 cans (about 1 lb. *each*) small whole onions
½ cup undiluted canned mushroom soup
½ cup shredded Swiss cheese
2 tablespoons Sherry or milk
¼ cup slivered or sliced almonds
1 tablespoon all-purpose butter

Drain onions and turn into a well greased 1-quart casserole. In a pan combine the mushroom soup, cheese, and Sherry; heat until cheese is melted. Pour over onions in the casserole. Either in the oven or in a small pan, heat and toast almonds in the butter until lightly browned. Sprinkle over top of onions. (If you prepare casserole ahead, put almonds on just before heating.) Bake, uncovered, in a 350° oven for about 25 minutes, or until heated through and bubbly. Cover dish immediately, and bring to table. Makes about 6 servings.

# Onion Soup

2 pounds yellow onions (7 or 8 of medium size)
2 tablespoons salad oil
3 tablespoons all-purpose flour
1 teaspoon bottled brown gravy sauce (optional)
2 quarts regular strength beef broth
½ cup dry white wine
½ teaspoon black pepper
½ teaspoon Worcestershire
1½ cups shredded Swiss cheese
¼ cup Sherry
6 large slices French bread, toasted

Cut onions into 1/4-inch-slices. Place in a 5-quart pan with oil, cover, and cook about 10 minutes. Uncover and continue to cook over moderate heat, stirring often, until lightly browned. Sprinkle flour over and stir. Add brown gravy sauce, the beef broth, dry white wine, pepper, and Worcestershire. Cover and simmer at least 30 minutes. Taste and correct seasonings. Stir in 1/2 cup of the Swiss cheese and the Sherry.

Pour soup into 6 individual ovenproof dishes (about 1 1/2 cups soup in each). In each dish, place a single slice of toasted French bread, trimmed as necessary to fit in the dish; cover slices with remainder of the cheese (about 2 1/2 tablespoons on each slice). Place under preheated broiler about 3 inches from heat for about 2 minutes to brown the bread and melt the cheese. Makes 6 servings.

## Tarte à L'Oignon

Serve as a first course at the table or as an appetizer.

  Pie crust mix for a 2-crust pie
½ cup dry beans
3 pounds yellow onions (about 8 medium-sized)
4 tablespoons (⅛ lb.) butter or margarine
1 teaspoon salt
3 tablespoons all-purpose flour
½ teaspoon ground cumin
½ cup half-and-half (light cream)
2 egg yolks, slightly beaten

Prepare the pie crust mix according to package directions. Press pastry together and roll out on a lightly floured board to a 13 by 18-inch rectangle. Fit pastry into bottom and sides of a 10 1/2 by 15 1/2-inch shallow baking pan. Prick pastry with fork, line with foil, and sprinkle with dry beans to hold in place. Bake in a 400° oven for 5 minutes; discard foil and beans, then bake 5 minutes more; cool.

Meanwhile, cut the onions in half lengthwise and peel; then slice halves thinly and evenly. Melt butter in a large frying pan with tight-fitting lid or a Dutch oven over medium heat. Add onions and salt, cover, and cook 15 minutes. Remove cover and continue to cook, stirring, until moisture is evaporated (about 15 minutes more). Stir in flour and cumin; gradually add half-and-half and bring to a boil, stirring. Remove from heat and stir in egg yolks.

Pour the filling into prepared crust. Bake in a 350° oven for 30 minutes; then broil until top is lightly browned (about 2 minutes). Cool slightly, then cut into 3 by 4-inch squares. Makes 12 servings.

## Onion Soufflé

  About 3 large onions
3 tablespoons butter or margarine
2 tablespoons all-purpose flour
½ teaspoon salt
⅛ teaspoon pepper
¼ cup half-and-half (light cream)
6 eggs, separated
2 tablespoons chopped pecans

Finely chop the onions (you should have 3 1/2 cups). Melt butter in a pan over medium heat; add onions, cover, and cook until onions are limp and liquid evaporated (about 20 minutes). Add the flour, salt, and pepper; stir until well blended. Remove from heat and gradually stir in cream. Return to heat and cook, stir-

ring, until thick. (This much can be done ahead; cover and refrigerate.)

If you cooked the onions ahead, stir over low heat until steaming. Remove from heat and stir in egg yolks. In a bowl, beat the egg whites until they hold short, distinct, moist-looking peaks. Fold beaten whites gently into onion mixture.

Pour into a well buttered 1 1/2-quart soufflé dish or casserole with foil collar, a 2-quart casserole, or 6 small (8 oz. size) custard cups. Sprinkle with nuts. Bake large soufflés in a 325° oven for 35 to 40 minutes (or in a 375° oven for 30 to 35 minutes) until center is firm when tapped. (Remove collar after 20 to 25 minutes or when set.) Bake small soufflés in a 350° oven for 20 to 25 minutes. Serve immediately. Serves 6 to 8.

## Danish Caramelized Onions

2 dozen boiling onions
  Boiling salted water
3 tablespoons butter or margarine
6 tablespoons firmly packed brown sugar

Peel onions and cook in gently boiling salted water for 10 minutes or until almost tender. Drain off water, and dry onions slightly on absorbent paper. In a large heavy frying pan slowly melt butter with brown sugar. Cook over low heat, stirring, until mixture is thoroughly blended. Add onions and cook, turning them in the syrup, until tender (about 10 minutes). Serves 6.

---

### *Cutting and Cooking Leeks*

Leeks require drastic trimming to separate the fibrous leaves from tender pale green parts. Also because of the way they grow, dirt has a way of lurking between the layers.

**To trim and clean leeks,** cut off root ends. Trim tops by making a diagonal cut from each side to a center point so only about 1 1/4 inches of dark green leaves remain. Strip away outer 2 or 3 layers of leaves until you reach the non-fibrous interior leaves. Split leeks in half lengthwise, hold each under running water and gently separate layers with your fingers to wash.

### Leeks Sautéed with Cream

Trim and clean 2 pounds (about 6) leeks as described under *cutting and cooking* above. In a large frying pan with a cover, cook leeks, covered, in about 4 cups boiling salted water until tender (about 10 to 15 minutes). Remove leeks from pan as each becomes tender; drain and discard cooking water. Cut leeks in 1-inch-long pieces; return to pan. Add 1 tablespoon butter, 3 tablespoons whipping cream, and 1/2 teaspoon crushed basil leaves. Boil over high heat, stirring gently, until liquid is evaporated. Makes 4 to 6 servings.

### Braised Leeks

Trim and clean 6 medium-sized leeks as directed above; drain. In a frying pan sauté 1 medium-sized cooking onion in 2 tablespoons butter or margarine until limp. Arrange sautéed onions evenly over bottom of frying pan or distribute in a shallow casserole. Arrange leek halves over top of onions. Cut 2 carrots in slanting slices about 1/2 inch thick and distribute over leeks. Pour over 1/2 cup regular strength chicken broth, sprinkle lightly with salt and pepper, and top with 2 bay leaves. Cover and simmer gently on top of the range or bake in a 350° oven about 40 minutes. Serves 6.

## Leeks and Tarragon

2 pounds (about 6) leeks
4 cups salted water
2 tablespoons butter or margarine
1 tablespoon chopped parsley
¼ teaspoon tarragon leaves, chopped
¼ cup grated Parmesan cheese

Trim and clean leeks following directions under *cutting and cooking*, page 66. In a large frying pan with a cover, bring water to boil; add leeks and simmer, covered, until tender (about 10 to 15 minutes), depending on sizes. Lift leeks from water as each is cooked tender; drain and arrange in a serving dish and keep warm. Melt butter and stir in parsley, tarragon, and cheese. Pour over leeks and serve. Makes 4 to 6 servings.

## Leek and Bacon Pie

8 or 10 slices bacon, chopped
3 cups thinly sliced leeks, trimmed and cleaned as directed under *cutting and cooking*, page 66
¼ cup water
1 teaspoon caraway seed
2 eggs
1 cup half-and-half (light cream)
    About ½ teaspoon salt
    Unbaked 9 or 10-inch pastry shell

Cook bacon until crisp, remove from pan, and set aside. Discard all but 3 tablespoons of drippings. Put leeks, water, and caraway seeds in pan; cook over high heat, stirring, until water evaporates and leeks are tender and slightly browned. Beat eggs with cream; add bacon, the leek mixture, and salt to taste. Pour into pastry shell. Bake on lowest rack in a 400° oven for 25 minutes or until crust is browned. Let stand 10 minutes before serving hot, or cool and serve. Makes 6 to 8 servings.

## Green Vegetable Soup

2 bunches leeks (6 to 8 medium-sized)
3 tablespoons butter or margarine
2 small potatoes, peeled and diced (about 1½ cups)
2 cans (about 14 oz. *each*) regular strength chicken broth
1 cup frozen peas, thawed
½ cup coarsely chopped watercress leaves
⅓ cup sour cream
    Salt and pepper
    Whole watercress leaves for garnish

Trim and clean leeks following directions under *cutting and cooking*, page 66. In a 3-quart saucepan melt the butter over medium heat; add leeks and cook until soft but not browned. Add potatoes and broth; bring to a boil, reduce heat, and boil gently, uncovered, for about 20 minutes, or until potatoes are almost tender. Add peas and cook 5 minutes longer. Stir in the chopped watercress.

Pour about 1/3 of the mixture at a time into a blender container and whirl until smooth; repeat until all the mixture is puréed. Return the soup to the saucepan, stir in the sour cream, and season to taste with salt and pepper. Reheat (do not boil). Garnish with a few whole watercress leaves. Makes 4 servings.

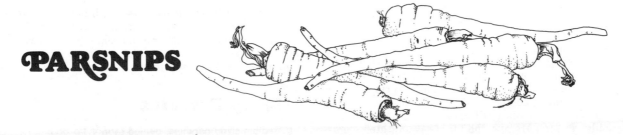

# PARSNIPS

### Selection and Storage

**Selection:** Choose small to medium-sized parsnips that have smooth, firm, well-shaped roots.
**Buying:** A pound of parsnips serves about 4.
**Storage:** Wrap, unwashed, in a plastic bag and keep in refrigerator. Will hold 1 to 2 weeks.

### Cutting and Cooking

The natural tapering shape of parsnips makes them a little tricky to cut in uniformly sized pieces that will cook evenly. You want them to be just tender throughout but not overcooked (they do fall apart and become mushy if overcooked). When ready to use, trim off tops and root ends, and scrape or pare the parsnips. They can be thinly sliced, diced, or shredded and cooked very quickly to uniform tenderness. For larger pieces, it's a good idea to cut parsnips crosswise into sections, then slice each section through the core to make pieces that are all the same thickness.

**To cook parsnip slices,** chunks, or diced parsnips,

put into boiling salted water. Cover and cook just until tender (about 5 to 10 minutes); drain immediately. (If done ahead, cool quickly in cold water, drain, and refrigerate.)

## Seasoning and Serving

Parsnips that have been sliced or diced and cooked quickly in boiling water can be seasoned in a variety of ways. A little sugar or honey heightens their naturally sweet flavor. Piquant or tart flavors—such as Worcestershire or lemon—are also good accents for parsnips. Here are some quick ways to season cooked parsnips. Each cup of parsnips makes about 2 servings.

### Parsnips in Cream

Cook diced parsnips as directed under *cutting and cooking*, page 67. For each 2 cups parsnips, blend in 1/4 cup whipping cream and 1/2 teaspoon Worcestershire. Heat, stirring, to simmering.

### Parsnips with Sour Cream

Cook diced parsnips as directed under *cutting and cooking*, page 67. For each 2 cups parsnips, blend in 1/3 cup sour cream and 1 1/2 tablespoons brown sugar (firmly packed). Heat, stirring, just to simmering.

### Marinated Parsnips

Cook 3 cups diced parsnips in 1/2 cup boiling water just until tender (5 to 10 minutes). Without draining parsnips, add 3/4 teaspoon basil leaves, 3 tablespoons white vinegar, 6 tablespoons salad oil, and salt and pepper to taste. Refrigerate until chilled. Serve on lettuce if desired.

---

### Butter-Steamed Parsnips

Melt 4 tablespoons butter or margarine in a wide frying pan or in an electric frying pan over highest heat. Add 4 cups thinly sliced parsnips and 6 tablespoons water. Cover and cook, stirring occasionally, for 6 minutes or until tender, adding more water if needed. Season with salt and pepper. Makes 4 to 6 servings.

### Spiced Parsnips

Prepare parsnips as directed above for Butter-Steamed Parsnips. Add 1/2 teaspoon caraway seed during the last 3 minutes of cooking time. Makes 4 to 6 servings.

# Stir-Fried Parsnips

3 to 4 cups sliced parsnips
  Boiling salted water
1 tablespoon salad oil
1 to 2 teaspoons finely chopped fresh ginger (optional)
1 to 2 cloves garlic, minced or mashed
1 tablespoon water
2 tablespoons brown or granulated sugar
  About 1/2 teaspoon salt
  About 1 tablespoon toasted sesame seed

Precook parsnips in the boiling water until barely tender when pierced with a fork. Drain, cool quickly with cold water, drain, and set aside.

Before starting to cook, prepare all other ingredients and have them within reach of your range. Heat a 10-inch or larger frying pan or a wok over high heat, then put in oil. As soon as oil is hot enough to ripple when the pan is tipped, add ginger (if used) and garlic; quickly stir with a spatula until browning starts (about 30 seconds). Put in parsnips, the 1 tablespoon water, sugar, and salt. Stir and fry about 1 to 2 minutes. Turn out on a warm serving dish and sprinkle with sesame seed. Makes 4 or 5 servings.

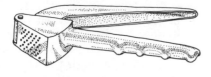

# Parsnip Carrot Sauté

1/2 pound *each* parsnips and carrots
3 tablespoons butter or margarine
4 tablespoons water
1/2 teaspoon dry mustard
1 1/2 teaspoons Worcestershire

Scrape or pare parsnips and carrots. Cut in slanting slices, graduating thickness of slices from 1/8 inch thick at the stem end to about 1/2 inch thick at the root end.

In a 10-inch frying pan over high heat, melt butter; add parsnips, carrots, water, mustard, and Worcestershire to the pan. Cover and cook, stirring occasionally until vegetables are tender when pierced (about 4 minutes). Makes 4 servings.

# Honey Parsnips

6 medium-sized parsnips, peeled (about 1 1/2 pounds)
  Boiling salted water
1 teaspoon salt
4 tablespoons (1/8 lb.) butter or margarine
4 tablespoons honey
  Ground cinnamon

Cut parsnips into pieces about 1 1/2 inches long. Then slice each through core into lengthwise slices about 1/2 inch thick. Cook in the boiling water just until tender (about 5 to 10 minutes); drain. Melt butter in a large frying pan; add honey, and cook, stirring, until bubbly. Add parsnips, cover, and cook over low heat for 5 minutes. Just before serving, sprinkle with cinnamon. Makes about 6 servings.

## Lemon Sautéed Parsnips

1 pound parsnips
  Boiling salted water
4 tablespoons (⅛ lb.) butter or margarine
1 tablespoon lemon juice
2 teaspoons honey
1 tablespoon minced parsley

Scrape or pare parsnips and cut crosswise into sections about 1 inch long. Cut each section through the core into 1/2-inch-thick chunks. Drop into boiling salted water and cook until parsnips are just tender when pierced (about 5 to 7 minutes); gently drain. (If done ahead, cool in cold water, drain, and refrigerate.)

Before serving, melt butter in a 10-inch frying pan. Stir in lemon juice, honey, and parsnips and cook over medium-high heat until glazed and lightly browned (about 10 minutes), turning gently. Sprinkle with minced parsley. Serves 4.

## Baked Glazed Parsnips

4 tablespoons (⅛ lb.) butter or margarine
1 pound parsnips
⅓ cup firmly packed brown sugar
¼ teaspoon salt
1 teaspoon grated lemon peel

Put butter in an 8-inch-square baking pan or casserole and place in a 375° oven to melt. Meanwhile, scrape or pare parsnips and cut crosswise into 3 to 4-inch lengths. Cut pieces lengthwise through the core so that the resulting pieces are about 1/2 inch thick.

Remove casserole from oven and stir in brown sugar, salt, and lemon peel. Add parsnips and turn in glaze. Return to oven and bake uncovered, stirring occasionally until parsnips are tender when pierced and most of liquid is absorbed (about 45 to 50 minutes). (Or bake in a 325° oven for about 1 hour until liquid is absorbed.) Makes 4 servings.

 **PEAS**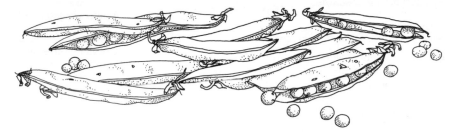

### Selection and Storage

**Selection:** Choose well-filled, bright, fresh, crisp green pods.
**Buying:** Allow about 3/4 pound peas in shells for each serving. 1 pound peas in shells yields about 1 cup shelled peas.
**Storage:** Store, unshelled, in refrigerator. Use as soon as possible.

### Cutting and Cooking

Shell peas from pods and rinse in cool water.
**To cook peas,** drop into about 1-inch boiling salted water. (If desired, add about 1 teaspoon sugar to the water.) Cover and simmer until just tender, but still bright green (5 to 10 minutes); drain. If you cook peas ahead, cool them quickly in cold water, drain, and refrigerate.

### Seasoning and Serving

Tender, young peas, freshly cooked, are naturally sweet and delicious seasoned only with butter, salt, and pepper. For variety, season with basil, mint, or tarragon leaves, chives, or ground nutmeg. For the following quick ways to season peas use about 3 cups shelled fresh peas or frozen peas.

#### Peas with Pine Nuts

Melt 4 tablespoons butter in a 10-inch frying pan. Add 1/2 cup pine nuts (or slivered almonds). Cook, stirring until nuts are golden. Add about 3 cups cooked peas and 1 clove garlic (minced or mashed); cook, stirring over high heat until heated through (about 3 minutes). Season to taste with salt and pepper.

#### Peas in Tangarine Sauce

In a pan blend 1 tablespoon cornstarch with 1 tablespoon wine vinegar. Add 3/4 cup tangarine juice, 1 tablespoon butter, and 3/4 teaspoon seasoned salt. Cook, stirring, until thick; pour over about 3 cups hot, cooked peas in a serving dish. Garnish with fresh tangarine sections.

#### Peas with Dill Butter Sauce

In a small pan heat 3 tablespoons finely chopped dill pickle in 1/4 cup (1/8 lb.) butter or margarine. Pour over about 3 cups hot, cooked, peas, seasoned to taste with salt and pepper.

#### Peas with Lemon Mint Butter

Cream together 1/2 cup soft butter, 1 tablespoon lemon juice, 1/4 teaspoon grated lemon peel, and 2 to 3 tablespoons finely chopped fresh mint. Melt the flavored butter generously over hot, cooked peas just before you serve.

#### Peas with Mushrooms

In a frying pan, sauté about 1/2 pound mushrooms, sliced, in 3 tablespoons butter or margarine for about 5 minutes. Add about 3 cups hot cooked peas and 2 to 3 tablespoons chopped pimiento. Mix lightly and serve.

## Butter-Steamed Peas

In a wide frying pan or electric frying pan on high heat, melt 2 tablespoons butter or margarine. Add 4 cups freshly shelled peas (or frozen peas, thawed) and 4 tablespoons water. Cover and cook, stirring frequently, for 3 minutes (2 minutes for frozen peas). Season with salt. Serves 5 to 6.

## Butter-Steamed Peas with Bacon

Follow the basic directions for Butter-Steamed Peas at left, making this change. Cook 4 slices coarsely chopped bacon in pan until browned; remove bacon with a slotted spoon and set aside. Add to drippings 4 cups freshly shelled peas and 6 tablespoons water.

Cover and cook, stirring frequently, for 3 minutes. Return bacon to pan. Season lightly with salt and pepper. Makes 5 to 6 servings.

# Sherried Peas and Mushrooms

2 cans (3 or 4 oz. *each*) sliced mushrooms
2 tablespoons butter or margarine
¼ teaspoon marjoram leaves
⅛ teaspoon ground nutmeg
2 tablespoons Sherry
2 packages (10 oz.) frozen tiny peas

Drain mushrooms, reserving liquid. In a frying pan, heat mushrooms in melted butter until sizzling, then stir in marjoram, nutmeg, and Sherry. Break peas apart and pour into pan with mushrooms; turn off heat and let stand. Just before you are ready to serve, add 2 tablespoons of the reserved mushroom liquid to peas and bring to boiling, stirring occasionally. Makes 6 servings.

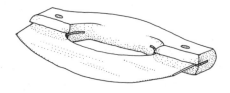

# Green Pea and Cheese Salad

¼ cup water
½ teaspoon *each* salt and sugar
1 package (10 oz.) frozen green peas
¼ pound finely diced American or Cheddar cheese (about 1 cup)
½ cup finely sliced celery
2 tablespoons chopped green onion, including part of tops
¼ cup thinly sliced radishes
3 tablespoons chopped sweet pickle
⅓ cup mayonnaise
Lettuce

Bring the water to a boil in a pan; add the salt, sugar, and frozen peas. Cover pan and bring to a boil over high heat; reduce heat and simmer 3 to 4 minutes. Turn into a bowl with ice cubes in the bottom. Cover and refrigerate until chilled. Drain peas well; add cheese, celery, green onion, and radishes. Combine the pickle and mayonnaise and stir into salad. Cover and chill about 1 hour. Serve on individual crisp lettuce leaves. Makes 4 to 6 servings.

# Pasta with Peas

1 package (12 oz.) frozen filled pasta (such as tortellini or raviolini, or fresh egg noodles such as tagliarini)
3 to 4 quarts boiling salted water
3 cups shelled peas (about 3 lbs. peas in shells) or frozen tiny peas, thawed
2 tablespoons butter or margarine
⅛ teaspoon freshly grated nutmeg
1½ to 2 cups whipping cream
1 egg, beaten
1 cup freshly grated or shredded Parmesan cheese
Additional grated or shredded Parmesan cheese and grated nutmeg

Drop filled pasta or noodles into rapidly boiling water over high heat; do not cover pan. When water resumes boiling, cook filled pasta 10 minutes or noodles 2 minutes, then add the fresh peas and cook 5 minutes longer (if you use frozen peas, cook only 2 minutes but allow 15 minutes total for filled pasta and about 7 minutes total for noodles). Drain peas and pasta.

In a wide frying pan melt butter, then add nutmeg and 1 1/2 cups of cream. Add peas and pasta and bring to a rapid boil.

Remove pan from heat and stir in the egg evenly; then mix in the 1 cup cheese. If the mixture is too thick, add more cream to smooth sauce. Sprinkle with more cheese and a little nutmeg. Serve at once. Makes 4 main dish or 6 first-course servings.

# Minted Green Pea Bisque

1 tablespoon finely chopped onion
¼ cup shredded carrot
1 package (10 oz.) frozen peas
1 sprig fresh mint (about 3 inches long), crushed
1 teaspoon sugar
Dash nutmeg
2 cups regular strength chicken broth
1 cup milk or half-and-half (light cream)
Salt and pepper
Chopped fresh mint (optional)

In a saucepan combine onion, carrot, peas, mint sprig, sugar, nutmeg, and chicken broth; simmer until vegetables are very tender (about 25 minutes). Pour a small amount at a time into a blender container and whirl until smooth (or put through a food mill). Stir in the milk and season to taste with salt and pepper. Chill 6 hours or overnight. Stir well; serve in chilled mugs; and garnish with chopped mint, if desired. Serves 4.

## Peas Cooked in Lettuce

Outside leaves from iceberg lettuce
1 package (10 oz.) small frozen peas, partially thawed
½ teaspoon salt
⅛ teaspoon pepper
¼ teaspoon whole fresh thyme leaves or pinch of dried
  thyme leaves
1 teaspoon scraped or grated onion, or ½ teaspoon instant
  minced onion
2 tablespoons butter or margarine

Wash the lettuce leaves, shake off excess water, and use them to line the bottom of a heavy frying pan (one with a tight cover) or electric frying pan.

Empty a package of frozen peas; break any large pieces that are frozen together and spread the peas on the lettuce. Sprinkle peas with the salt, pepper, thyme, and onion; dot the top with butter. Cover the pan, turn the heat to medium, and boil just until the peas are tender (about 3 minutes). Turn peas into a serving dish, discard the lettuce, and serve immediately. Serves 3.

## Green Peas, French-Style

¼ cup (⅛ lb.) butter or margarine
1 small head lettuce, finely shredded
6 cups fresh or frozen peas (about 6 lbs. in shell, or 4
  packages, 10 oz. *each*)
8 sprigs parsley, tied
2 teaspoons salt
½ teaspoon sugar
⅛ teaspoon ground nutmeg
1 tablespoon *each* all-purpose flour and butter or
  margarine

In a large saucepan, melt the 1/4 cup butter, and put in lettuce. Make a layer of peas over the lettuce, then add parsley, salt, sugar, and nutmeg. Cover and cook over medium heat until peas are tender (about 10 minutes); stir once lightly.

Cream together the flour and 1 tablespoon butter. When peas are tender, stir flour mixture in lightly and return to heat until liquid is bubbly and slightly thick-

ened. Remove parsley. Serve immediately. Makes about 12 servings.

## Minted Pea Salad

1 medium-sized cucumber, peeled
2 packages (10 oz. *each*) frozen tiny peas, thawed and
  drained
3 tablespoons finely chopped fresh mint or ½ teaspoon
  crushed mint leaves
1 tablespoon lime or lemon juice
⅓ cup mayonnaise
1 tablespoon honey
½ teaspoon salt
  Lettuce leaves
  Mint sprigs (optional)

Cut cucumber in half lengthwise; use a spoon to scoop out and discard seeds. Coarsely chop cucumber and combine in a bowl with the peas and chopped mint. In a small bowl, stir together the lime juice, mayonnaise, honey, and salt until blended; pour dressing over peas and mix well. Serve on individual lettuce leaves. Garnish with mint sprigs. Makes 6 to 8 servings.

## Fried Rice with Peas

½ cup (¼ lb.) butter or margarine
½ cup sliced green onion, including part of tops
¼ cup minced parsley
1 package (10 oz.) frozen peas
  Boiling, salted water
4 cups cooked rice, hot or cold
2 teaspoons grated lemon peel
2 tablespoons soy sauce
  Dash liquid hot pepper seasoning

In a frying pan or wok, heat the butter; add green onion and parsley and sauté just until limp and bright green (about 3 minutes). Meanwhile, add peas to boiling water and bring just back to boiling; remove from heat, drain, and set aside. Add the rice, lemon peel, soy sauce, hot pepper seasoning, and peas to the pan. Stir over heat until blended and heated through. Cover pan and keep warm. Makes 4 servings.

# PEAS, EDIBLE POD

## Selection and Storage

**Selection:** These may also be called snow, sugar, or Chinese pea pods. Look for firm, crisp pods that are a bright green color.
**Buying:** One pound edible pod peas makes about 4 servings.
**Storage:** Seal edible pod peas inside a plastic bag to keep in the natural moisture and store in the refrigerator. Use as soon as possible.

## Cutting and Cooking

Snap off both ends and remove strings from these peas as you would green beans. Rinse before cooking. To preserve their sweet crispness and bright color, you must be careful not to overcook them.

**To cook edible pod peas,** remove ends and strings; drop into a large quantity of boiling salted water. Cook, uncovered, on high heat just until boil resumes. Drain and serve at once.

**Butter-Steamed Pod Peas**

In a wide frying pan or electric frying pan on high heat, melt 2 tablespoons butter. Add 3 cups edible pod peas (ends and strings removed) and 4 tablespoons water. Cover, cook, stirring frequently for 2 minutes. Season with salt and sugar.

**Cream-Glazed Pod Peas**

Follow directions at left for Butter Steamed Pod Peas with these few changes. After cooking for 2 minutes, remove frying pan cover, add salt to taste and 1/4 cup whipping cream. Cook, stirring until liquid is almost gone.

# Snow Pea Salad with Sesame Dressing

1 package (7 oz.) frozen edible pod peas
  Boiling salted water
½ head cauliflower
1 can (5 oz.) water chestnuts, drained and sliced
1 tablespoon chopped pimiento
  Sesame Seed Dressing (recipe follows)

Cook peas in a small amount of boiling salted water until barely tender (about 1 minute); drain. Separate cauliflower into bite-sized clusters (you should have about 2 cups); cook in boiling salted water until tender but still crisp (about 3 minutes); drain. Combine peas and cauliflower with water chestnuts and pimiento; cover and chill. Just before serving, mix with about 3 tablespoons of the Sesame Seed Dressing. Makes 4 to 6 servings.

**Sesame Seed Dressing.** Place 2 tablespoons sesame seed in a shallow pan in a 350° oven for 5 to 8 minutes or until golden brown; cool. In a covered jar, combine 1/3 cup salad oil, 1 tablespoon each lemon juice, vinegar, and sugar, 1/2 clove garlic, minced or mashed, 1/2 tea-spoon salt, and the toasted sesame seed. Cover and chill. Shake well before using.

# Skillet Snow Peas with Celery

½ cup sliced green onion tops
4 stalks celery, cut in thin slanting slices
1 pound edible pod peas, ends and strings removed
1 tablespoon cornstarch
½ teaspoon salt
1 teaspoon sugar
1 tablespoon soy sauce
½ cup water
2 tablespoons salad oil

Prepare the green onion, celery, and pod peas and have within reach of your range. In a small bowl, stir together the cornstarch, salt, sugar, soy sauce, and water; set aside. In a large frying pan or wok, heat salad oil over high heat. Add onion, celery, and peas; stir-fry 2 minutes. Add cornstarch mixture to vegetables, and cook, stirring, until peas are barely tender (about 2 minutes more). Serve immediately. Makes 4 to 6 servings.

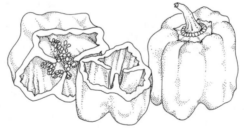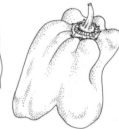

# PEPPERS
*Green, Red*

## Selection and Storage

**Selection:** Choose firm, well shaped peppers with thick brightly colored flesh. When green peppers are allowed to ripen fully, they turn brilliant red. These red bell peppers are generally available in late summer and fall.

**Buying:** Allow 1/2 to 1 whole pepper for each serving.

**Storage:** Ideal storage temperature is about 50°. If cool storage area is not available, place peppers, unwashed, in crisper section of refrigerator; use within one week.

## Cutting and Cooking

Cut peppers in half, remove stems, and rinse and cut away seeds and pith; then cut into slices, slivers, strips, or chop them. For stuffed whole peppers, slice off top, and rinse out seeds. Or cut peppers in half lengthwise or crosswise when you want pepper shells.

**To parboil peppers** when you plan to stuff and bake them, drop pepper shells or seeded whole peppers into a large quantity of boiling salted water. Boil for about 3 minutes; remove, rinse in cold water, and turn upside-down to drain.

## Seasoning and Serving

The green and red bell peppers can be used interchangeably in many recipes, however their flavors are distinctly different. When the peppers mature and turn red, they develop fuller, more mellow, and much sweeter flavor. These flavor differences are accentuated in the recipes for sautéed peppers that follow.

### Sautéed Green Peppers

Rinse and seed 3 green peppers and cut them in slivers. Heat 1 tablespoon butter in a wide frying pan, add peppers, and cook over high heat, stirring often, for about 5 minutes or until peppers are bright green and just beginning to lose their crispness. Serve hot or warm.

### Sautéed Red Peppers with Cumin

Rinse and seed 4 red bell peppers; cut in thin slices. Heat 3 tablespoons salad oil in a wide frying pan; add peppers, 2 tablespoons water and 1/2 teaspoon whole cumin seed. Cook over medium heat, stirring, until peppers are limp and liquid has evaporated; salt to taste. Serve at room temperature.

### Piquant Red Peppers

Rinse and seed 1 large red bell pepper; cut in slivers. Put into a frying pan with 2 tablespoons *each* olive oil and water. Cover and cook over high heat until liquid evaporates. Remove cover and stir in 1 tablespoon wine vinegar and salt to taste. Serve at room temperature. This makes a fine relish with toasted cheese and ham sandwiches.

### Italian-Style Peppers

Rinse and seed 4 large green peppers; cut into strips about 1 1/2 inches wide. Heat 2 tablespoons each olive oil and butter in a wide frying pan over medium heat. Add peppers and 1 clove garlic, minced or mashed. Cook, stirring occasionally, until lightly browned. Season with 1 1/2 teaspoons salt, 1/8 teaspoon pepper, and 1 teaspoon oregano leaves. Cover and cook over low heat for 15 minutes, or until tender. Makes about 4 servings as a vegetable accompaniment to meats.

### Baked Red Bell or Green Peppers

Cut lengthwise into quarters 3 large red (or green) bell peppers; rinse and remove seeds. Arrange pieces close together, skin side down, in a close-fitting baking pan (about 8 by 12 inches). Sprinkle with 1/4 cup seasoned crumbs (recipe follows); drizzle with 1/4 cup olive oil. Bake, uncovered, in a 375° oven for 45 minutes. Cool to room temperature. Serve with lemon wedges to squeeze onto each serving. Makes 4 to 6 servings as vegetable accompaniment to meats.

*Seasoned crumbs.* Cut end from 1 pound loaf sour or sweet French bread. Slice and toast enough of the center portion (about 1/2 loaf) to make 1 1/2 cups coarse crumbs; whirl covered in a blender. Mix crumbs with 1/2 cup minced parsley, 3/4 teaspoon salt, 1 clove garlic (minced or mashed), 1 teaspoon crumbled basil leaves, 1/4 teaspoon *each* crumbled rosemary and rubbed sage, and 6 tablespoons olive oil. Makes about 1 1/2 cups.

---

### Butter-Steamed Green Peppers

Chop 1 medium-sized onion. Melt 2 tablespoons butter in a wide frying pan or electric frying pan over medium heat. Add onion and cook until soft. Turn heat to high and add 3 large, seeded, sliced green peppers (about 4 cups) and 5 tablespoons water; cover; cook, stirring occasionally, for 6 minutes. Remove cover; salt to taste. Serves 6.

### Green Peppers and Onions with Cheese

Prepare green peppers and onion as directed above, adding 3/4 cup shredded sharp Cheddar cheese along with the salt. Stir until cheese is melted. Makes 5 to 6 servings.

# Red Pepper Scrambled Eggs

3 tablespoons butter or margarine
⅓ cup chopped onion
1 red bell pepper, seeded and diced
6 eggs
½ teaspoon salt
⅛ teaspoon pepper
3 tablespoons half-and-half (light cream)

Heat the butter in a frying pan. Sauté onions and peppers, stirring occasionally until tender (about 10 minutes). Meanwhile, beat together with a fork the eggs, salt, pepper, and cream.

Pour into pan and cook slowly, stirring often until eggs are scrambled and softly set. Makes 4 servings.

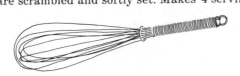

# Picnic Shrimp Salad

Pack the salad elements in a picnic basket to eat out of hands.

1 package (3 oz.) cream cheese, softened
¼ cup minced parsley
2 teaspoons wine vinegar
  About ⅓ cup sour cream
½ to ¾ pound small, whole, cooked shrimp
  Salt and pepper to taste
3 red (or green) bell peppers, rinsed
2 small heads romaine lettuce

In a bowl mash cream cheese and blend in parsley, wine vinegar, and enough sour cream to make an easily spooned mixture. Mix in shrimp; season to taste with salt and pepper.

To serve, cut peppers in big chunks and break romaine leaves. Spoon shrimp salad onto vegetables and eat. Makes 4 to 6 servings.

## Mediterranean Salad

2 tablespoons olive oil
1 medium-sized onion, finely chopped
1 medium-sized eggplant, peeled and cut in small cubes
1 teaspoon salt
½ cup uncooked rice
1 cup water
2 tablespoons catsup
2 green peppers
    Mayonnaise Sauce (recipe follows)

Heat the oil in a medium-sized saucepan. Add the onion and eggplant cubes; cook about 5 minutes or until limp. Add the salt, rice, and water. Cover and cook over low heat until the rice is tender. Remove from heat and stir in the catsup.

Cut the peppers in halves, remove and discard seeds, and fill with rice mixture. Place in a shallow baking dish. Pour about 1/4 cup water in the bottom of the pan. Bake, uncovered, in a 350° oven for 30 to 40 minutes or until the peppers are tender but firm. Chill. Serve cold topped with mayonnaise sauce. Makes 2 to 4 servings.

**Mayonnaise Sauce.** Mix together 1/2 cup mayonnaise with 2 tablespoons sweet pickle relish and 3 tablespoons lemon juice.

## Green Pepper Appetizers

4 large bell peppers
⅓ cup vinegar
¼ cup minced onion
½ teaspoon salt
¼ teaspoon pepper
½ cup olive oil or salad oil
1 clove garlic, minced
½ teaspoon basil leaves
1 crumbled bay leaf
1 package (6 oz.) smoked Cheddar cheese, shredded
⅓ cup melted butter
60 saltine crackers

Place peppers in a 400° oven until the skin puffs (about 10 to 12 minutes). Remove stems, seeds, and peel; cut in 8 lengthwise slices, then halve each slice crosswise. Combine vinegar, onion, salt, pepper, oil, garlic, basil, and bay leaf in a small bowl. Add pepper slices, and marinate overnight in the refrigerator.

Combine cheese and melted butter in a bowl, mix well, and chill. To serve, let guests spread cheese mixture on a cracker, then top with a piece of pepper. Makes about 60 appetizers.

## Pepper Shells with Potatoes

4 medium-sized potatoes, peeled, cooked, drained, cooled
6 short, fat, red bell or green peppers
6 tablespoons butter or margarine
1 medium-sized onion, chopped
½ teaspoon each ground coriander, ground cumin, and mustard seed
½ teaspoon ground turmeric or curry powder
1 teaspoon salt
⅛ teaspoon pepper
2 teaspoons lemon juice

Dice potatoes into 1/2-inch squares. Cut a slice off top of each pepper to make shells 2 1/2 inches high. Remove seeds, keeping shells intact. Then cut stem out of each top; discard; finely chop remaining pepper tops; set aside. Par-boil shells as directed under *cutting and cooking*, page 72; turn upside-down to drain.

In a large frying pan, melt butter over medium-high heat. Add onion, potatoes, and chopped pepper; cook, turning until golden brown (about 8 to 10 minutes). Add coriander, cumin, mustard seed, turmeric, salt, and pepper during the last few minutes. Remove from heat and sprinkle with lemon juice.

Arrange shells in a greased, shallow baking dish. Pile potatoes into shells. Bake, uncovered, in a 350° oven 20 minutes. Makes 6 servings.

## Corned Beef, Cabbage, Peppers

1 medium-sized head cabbage
1 large white or yellow onion
2 red bell peppers
2 cans (12 oz. each) corned beef
1 cup water
4 drops liquid hot pepper seasoning
2 teaspoons soy sauce
2 tablespoons vinegar
1 teaspoon sugar

Cut cabbage into about 1-inch-wide wedges. Slice onion in rings. Remove stems and seeds from peppers and slice in 1/4-inch-wide strips. Cut corned beef in thin slices. Using a Dutch oven or other heavy pan, arrange half the cabbage on the bottom; top with half the onion, peppers, and beef slices. Repeat layers using remaining cabbage, onion, peppers, and beef. Combine the water, hot pepper seasoning, soy, vinegar, and sugar; pour over foods in pan. Cover, bring to a boil, reduce heat, and simmer gently until tender (about 30 minutes). Serve in wide soup bowls or deep plates. Makes 6 to 8 servings.

## Green Pepper Casserole

2 tablespoons butter or margarine
1 cup uncooked long grain rice
3 chicken bouillon cubes
2 cups boiling water
3 green peppers, seeded and sliced
1½ pounds lean ground beef
1 medium-sized onion, chopped
2 stalks celery, thinly sliced
1 clove garlic, minced or mashed
1 teaspoon each oregano leaves and salt
¼ teaspoon pepper
2 cans (8 oz. each) tomato sauce
1½ cups shredded Cheddar cheese

Heat the butter in a frying pan, add rice, and stir until lightly toasted; spoon into a greased 9 by 13-inch baking pan. Dissolve bouillon cubes in boiling water and pour over rice. Arrange green pepper slices on the rice. Cover tightly with foil and put into a 375° oven for 20 minutes.

Using the same frying pan, sauté meat until it loses its pinkness. Add the onion, celery, garlic, oregano, salt, and pepper; continue cooking for 5 minutes. Stir in the tomato sauce. Remove casserole from oven and spoon mixture over peppers. Cover and bake for 15 minutes. Remove cover, sprinkle with cheese, and bake, uncovered, for 5 more minutes. Makes 6 servings.

## Savory Stuffed Green Peppers

6 medium-sized green peppers
1 teaspoon salt
1 pound lean ground beef
1 medium-sized onion, chopped
¼ teaspoon *each* garlic salt and ground cinnamon
1½ cups cooked brown rice
⅓ cup golden raisins
⅙ cup salted cashews, chopped
1 can (10 oz.) white sauce
½ cup shredded Cheddar Cheese

Slice off stem end of peppers and remove seeds; par-boil as directed under *cutting and cooking*, page 72. Cool in cold water; drain and set aside.

In a frying pan, heat the salt over medium heat; add crumbled beef and onion and cook, stirring, about 5 minutes. Remove from heat and add garlic salt, cinnamon, cooked rice, raisins, nuts, and about 1/2 cup of the white sauce; mix well and spoon into peppers. Stand peppers up in a shallow baking pan. Combine remaining white sauce and cheese and spoon equally over top of each pepper. Add 1/2 inch boiling water to pan and bake, uncovered, in a 350° oven for 30 minutes or until meat is hot. Makes 6 servings.

## Stuffed Peppers Guatemala

2 or 3 slices white bread, cubed
2 tablespoons salad oil
2 cups chopped onion
1 pound (about 3 cups) finely chopped fully cooked
    Canadian bacon or ham
1 can (8 oz.) tomato sauce
½ teaspoon oregano leaves, crumbled
¼ teaspoon thyme leaves, crumbled
¼ cup finely chopped parsley
4 medium-sized red or green peppers, cut in half
    lengthwise; seed and remove pith

Whirl enough of the bread in a blender to make 1 cup soft crumbs; set aside.

Heat salad oil in a 10-inch frying pan over medium high heat. Add onion and sauté until soft (about 10 minutes). Stir in bacon or ham, tomato sauce, the crumbs, oregano leaves, thyme leaves, and parsley. Boil until most of the liquid evaporates (about 5 minutes). Cover and cool to room temperature. Meanwhile, parboil peppers as directed under *cutting and cooking*, page 72; drain well. Set hollow-side-up in an ungreased baking pan.

Divide the meat mixture evenly among the pepper shells. Bake, uncovered, in a 375° oven for 35 minutes or until heated through. Use a wide spatula to transfer to a serving platter. Makes 4 servings.

## Marinated Red Pepper Salad

1 cucumber, peeled and thinly sliced
1 teaspoon salt
2 medium-sized red bell peppers
¼ pound medium-sized mushrooms
½ cup salad oil
⅓ cup white wine vinegar
1 teaspoon *each* basil leaves and chervil leaves, crumbled
¼ teaspoon seasoned salt
⅛ teaspoon pepper
    Lettuce

Put cucumber in a bowl; sprinkle with salt, cover, and chill. Discard stems and seeds from peppers; thinly slice in rings. Rinse mushrooms and slice lengthwise through the stems. Combine peppers and mushrooms in a bowl. Mix together the salad oil, vinegar, basil, chervil, seasoned salt, and pepper. Pour over peppers and mushrooms. Marinate for 1 to 2 hours.

To serve, arrange the cucumbers, peppers, and mushrooms on a bed of lettuce on the serving plate; spoon over a little of the marinade over top. Makes about 6 servings.

# POTATOES
*White, Sweet*

## Selection and Storage

**Selection:** For general purpose potatoes and baking potatoes, look for reasonably smooth skins that are free from blemishes, decay, root sprouting, or skinned surfaces. "New" potatoes should be well shaped, firm, free from blemishes and sunburn (a discoloration under the skin).

**Buying:** Allow 1 baking potato for each serving; 2 to 3 new potatoes for each serving.

**Storage:** If stored in a cool, dark, dry area (45 to 50° temperature is best) with good ventilation, baking potatoes keep several months and new potatoes for several weeks. Potatoes stored at room temperature should be used within a week.

## Cutting and Cooking

The type and maturity of potatoes often determines the ways to cook them. New potatoes, whether red and round or brown and oval, feel smooth and almost silky. When boiled, they have smooth, waxy texture. Baking potatoes are usually brown with slightly rough (or netted) skin; they are more fluffy and mealy when cooked. Often recipes call for one or another of these types when that particular texture is required for a potato dish. The many varieties of general purpose potatoes may also be baked or boiled. Or they may be cut into quarters, fingers, shoestrings, sliced, or diced to cook as directed in individual recipes.

**To bake potatoes,** wash and scrub the skin and pat dry. Rub skin of potatoes with butter, olive oil, bacon drippings, or salad oil. Pierce skin with tines of a fork to allow steam to escape while it bakes. Place on the oven rack in a 450° oven and bake until tender (about 40 minutes or longer depending on size). If you bake potatoes at lower temperature along with a roast, they will take about 1 1/2 hours or longer.

**For boiled potatoes,** use small whole new potatoes with skins scrubbed, or peeled larger potatoes cut in quarters. Cook, covered, in boiling salted water until just tender (20 to 30 minutes); drain.

**To oven roast potatoes** in the pan drippings alongside a roast, use small, whole, new potatoes or large potatoes cut in half lengthwise. About 1 hour before roast or poultry is done, roll new potatoes around in the drippings, then turn and baste occasionally as the meat cooks. For larger halved potatoes, cut 1/8-inch-deep lines in a crisscross design on cut surface. Place cut side down in drippings and put in about 1 1/2 hours before roast is done. After about 1 hour, turn over and baste.

### Seasoning and Serving

Boiled potatoes may be served plain to season with butter. Or you might dress the drained potatoes with melted butter to which any of these seasonings has been added: chopped chives, minced parsley, equal amounts of chopped parsley and mint, caraway seed, grated Parmesan cheese, toasted bread crumbs, crisp bacon bits, curry powder, or dill weed.

### Butter-Browned New Potatoes

Peel 3 pounds uniformly small new potatoes; wash and dry. Melt 3/4 cup butter or margarine in a Dutch oven over medium heat. Put in potatoes—they should just cover pan bottom. Cook, uncovered, for 15 minutes, stirring carefully. Reduce heat to low, cover pan, and cook 10 to 15 minutes or until tender. Uncover, increase heat to medium-high, and keep turning potatoes until crisped and browned. Lift out with slotted spoon. Sprinkle with salt, pepper, and chopped parsley. Makes 12 servings.

### Butter-Roasted Diced Potatoes

Boil 2 pounds large new potatoes following directions for boiled potatoes under *cutting and cooking* above; drain, cool, peel, and finely dice. Melt 5 tablespoons butter or margarine in a 10 or 11-inch heavy frying pan; add potatoes and 1/4 teaspoon salt; stir to distribute butter evenly. Press potatoes into an even layer in the pan. Sprinkle with 2 teaspoons water, cover tightly, and cook over medium-low heat until crusty and golden at the bottom (about 15 minutes). Slip spatula under potatoes to loosen all the way around. Invert onto serving plate. Makes about 6 servings.

## Dollar Potatoes

    4 large potatoes
    ½ cup butter
    2 green onions, including part of tops, finely chopped

Scrub potatoes, leaving on skins. Cut crosswise in about 3/16-inch slices, keeping potato shape intact. Arrange in 2 rows, with slices upright, in a buttered 5 by 9-inch loaf pan. Melt butter, add onions, and pour over potatoes. Bake, uncovered, in a 425° oven for 1 hour; reduce heat to 375° and bake 30 minutes longer or until tender. Turn out of pan onto a platter to serve. Makes 8 to 10 servings.

## Oven French Fries Parmesan

    3 large potatoes
    Ice water
    ¼ cup (⅛ lb.) butter or margarine
    Onion salt *or* garlic salt and paprika
    ¼ cup Parmesan cheese, grated

Scrub potatoes, do not peel; cut into sticks as for French fries. Cover with ice water; allow to stand 30 minutes. Drain and dry. Melt butter in 2 baking pans (about 10 by 15 inches each). Add potatoes and toss until all sides are butter-coated. Spread out in a single layer. Sprinkle to taste with onion salt or garlic salt and paprika. Bake in a 450° oven 25 minutes or until tender and brown. Turn occasionally. Remove from oven, sprinkle with cheese, shaking pan so potatoes are evenly coated. Makes 4 generous servings.

## Potatoes with Fresh Tomatoes

    4 medium-sized potatoes
    2 medium-sized firm ripe tomatoes
    1 medium-sized onion
    ½ green pepper
    2 tablespoons shortening or salad oil
    4 tablespoons (⅛ lb.) butter or margarine
    ½ teaspoon salt
    ⅛ teaspoon pepper
    ½ teaspoon sugar

Peel the potatoes and cut into 1/2-inch cubes. Peel the tomatoes, cut in half and gently squeeze out the seed pockets; dice. Peel and thinly slice onion and seed the green pepper and thinly slice.

In a heavy frying pan or electric frying pan, heat

the shortening and 2 tablespoons of the butter over medium heat (about 350° on electric pan). Sauté the potatoes slowly, turning over as needed until golden and tender (about 25 minutes); sprinkle with salt and pepper.

Meanwhile, in a small pan, sauté the onion, green pepper, and tomato in remaining 2 tablespoons butter until soft but not brown; sprinkle with the sugar and keep warm. Turn the potatoes into a warm serving dish and spoon the tomato mixture over the top. Serve immediately. Makes 6 to 8 servings.

## Potato Salad Casserole

8 slices bacon
1 small onion, finely chopped
⅓ cup vinegar
1 teaspoon salt
¼ teaspoon pepper
4 teaspoons sugar
2 cans (1 lb. *each*) small potatoes
½ green pepper, seeded and finely chopped

Cook the bacon until crisp; remove from the frying pan and drain. Leave the bacon drippings in the pan; add the onion, vinegar, salt, pepper, and sugar to drippings in frying pan. Cook, uncovered, for about 3 minutes.

Slice the potatoes into a 1 1/2-quart casserole. Crumble in the bacon slices, then add the chopped green pepper. Pour in the hot dressing and mix together lightly.

Bake, uncovered, in a 375° oven for 20 minutes. Serve hot. Makes about 6 servings.

## Fluffy Potato Casserole

2 cups hot or cold mashed potatoes
1 large package (8 oz.) cream cheese, at room temperature
1 small onion, finely chopped
2 eggs
2 tablespoons all-purpose flour
   Salt and pepper to taste
1 can (3½ oz.) French fried onions

Put the potatoes into the large bowl of your electric mixer. Add the cream cheese, chopped onion, eggs, and flour. Beat at medium speed until the ingredients are blended, then beat at high speed until light and fluffy. Taste and add salt and pepper, if needed. Spoon into a greased 9-inch-square baking dish. Distribute the canned onions evenly over the top. Bake, uncovered, in a 300° oven for about 35 minutes. (If you prepare this dish ahead, add the onions just before putting it in the oven.) Makes 6 to 8 servings.

## Potato Parmesan Soufflé

8 servings instant mashed potatoes, prepared as per package directions
6 eggs, separated
¾ cup shredded Parmesan cheese
½ teaspoon *each* salt and cream of tartar

Prepare instant mashed potatoes following package directions for 8 servings. Place in a bowl and while still hot, beat in egg yolks, one at a time, then mix in cheese.

(At this point, you can cover and let stand at room temperature a few hours.) About 1 hour before serving time, beat egg whites until foamy, add salt and cream of tartar, and beat until soft peaks form. Fold beaten whites into potato mixture. Spoon into two buttered 2-quart soufflé dishes or a 4-quart soufflé dish (or use baking dishes with straight sides).

Bake in a 375° oven allowing 45 minutes for small soufflés or 1 hour for a large soufflé. Makes 8 to 10 servings.

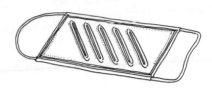

## Rancher's Potato Boats

6 medium-sized baking potatoes, baked and warm
3 tablespoons butter or margarine
½ to ¾ cup milk
½ cup *each* sour cream and shredded Parmesan cheese
10 slices bacon, cooked and crumbled
½ cup chopped green onions, including part of tops
   Sour cream for topping (about 1 cup)

Cut a thick lengthwise slice from each potato; scoop hot potatoes into a bowl, reserving shells.

With a potato masher or wire whip, beat in the butter, milk, sour cream, and cheese. If necessary, add additional milk to make a soft, fluffy consistency. Mix in bacon and onions. Pile into potato shells, mounding tops. Bake in a 425° oven for 15 minutes or until hot through and browned. Pass additional sour cream to spoon on top. Makes 6 servings.

## Potato Pancakes

4 medium-sized baking potatoes, peeled
1 teaspoon ascorbic acid mixture
1 tablespoon grated onion (optional)
2 eggs, slightly beaten
2 tablespoons all-purpose flour
1 teaspoon salt
   Dash pepper (optional)
¼ teaspoon baking powder
   Peanut or vegetable oil

Shred potatoes on a fine shredder (you should have about 3 cups). Add ascorbic acid mixture to first potato you have grated, and stir to incorporate it as more are added. Blend with onion and eggs. Mix together flour, salt, pepper, and baking powder; blend into potato mixture. Pour oil to a depth of about 1/8 inch in a heavy frying pan or electric frying pan set at 370°; heat until very hot.

Using about 1/3 cup batter for each pancake, drop potato mixture into the hot fat. Brown on both sides over moderately-high heat, turning only once. Serve pancakes crisp and hot. If necessary, keep pancakes warm on a cooky sheet lined with paper towels in a warm oven; cover with a sheet of foil with holes punched in it with a fork to release steam. Makes about 10 pancakes 4 inches in diameter.

# German-Style Potato Salad

2 pounds (6 medium-sized) white new potatoes
  Boiling water
1 small mild, red onion
10 slices bacon
1 egg, slightly beaten
½ cup sugar
1 tablespoon all-purpose flour
½ teaspoon *each* dry mustard and salt
⅓ cup *each* water and white wine vinegar
  Salt and pepper to taste

Cook potatoes in boiling water until just tender when pierced (about 30 minutes); drain. When cool enough to handle, peel and cut into 1/2-inch cubes. Thinly slice onion, separate into rings, and add to potatoes. Cook bacon until crisp; drain and crumble 7 slices into the potatoes. Set aside remaining bacon for garnish.

Put egg in a saucepan and stir in the sugar, flour, mustard, salt, water, and white wine vinegar. Cook over medium heat, stirring, until mixture thickens (about 4 minutes). Pour over potatoes; mix well and season with salt and pepper to taste. Cool, cover, and chill at least 4 hours or overnight. Before serving, garnish with reserved bacon, crumbled. Makes 6 to 8 servings.

# Spinach-Filled Baked Potatoes

4 medium-sized baking potatoes (about 2 lbs.), baked and warm
½ cup soft butter or margarine
1½ teaspoons salt
1 teaspoon *each* sugar and dill weed
¼ teaspoon pepper
1 package (10 oz.) frozen, chopped spinach, thawed
¼ cup chopped chives (fresh, frozen, or freeze-dried)
¼ cup grated Parmesan cheese

Slit each potato lengthwise down the center and carefully scoop out inside, leaving the potato skin shell. Set the shells aside. Place potatoes into the large bowl of an electric mixer and add the butter, salt, sugar, dill, and pepper. Beat at medium speed until the ingredients are blended and the potatoes are mashed smoothly. Squeeze excess moisture from spinach and add with chives to potatoes, blending well. Refill shells with mashed potato mixture. (If made ahead, wrap and chill at this point.) Bake, uncovered, in a 350° oven for about 30 minutes or until heated through (allow 45 minutes for cold potatoes). Remove from oven and immediately sprinkle each potato with an equal amount of the cheese. Makes 4 servings.

# Tuna-Potato Pie

1 can (1 lb.) whole potatoes, sliced
2 cans (6½ oz. *each*) chunk-style tuna, thoroughly drained
6 ounces Swiss cheese, shredded (about 2 cups)
½ cup sliced green onions, including tops
4 eggs
2 cups half-and-half (light cream)
1 teaspoon dill weed
½ teaspoon salt
⅛ teaspoon pepper

Lightly grease a 1 1/2-quart shallow casserole and arrange potato slices in an even layer on the bottom;

cover with tuna (flaking from can with fork), then sprinkle in half the cheese and half the onions. Beat eggs until blended; stir in half-and-half, dill, salt, and pepper and pour carefully into casserole. Sprinkle remaining cheese evenly over surface and bake in a 325° oven for 45 minutes or until center jiggles slightly when dish is shaken (cooking times will vary according to pan size; begin checking pie at about 35 minutes). Sprinkle remaining onions in a band down the center and serve. Makes 6 servings.

# Curried Potato Patties

2 cups hot mashed potatoes
1 egg
¼ teaspoon curry powder
1 teaspoon chopped parsley
1 tablespoon butter or margarine

Place the hot mashed potatoes in a mixing bowl, add the egg, curry powder, and parsley. Mix until well blended. On a greased baking sheet, shape the mixture into 4 patties and flatten slightly in the centers. Put a dot of butter in the center of each patty. Bake in a 400° oven for 10 minutes or until heated through. Makes 4 servings.

# Roast Beef Hash

1½ quarts finely diced (¼-inch cubes) raw potatoes
  About ½ cup all-purpose flour
  About 2 tablespoons *each* bacon drippings and salad oil or butter
2 medium-sized onions, very finely diced
  Salt and pepper to taste
1½ quarts medium diced (about ¼-inch cubes) cooked roast beef (do not grind)

Drain potato cubes on absorbent material, then sprinkle with flour.

Heat bacon drippings and oil in a large frying pan. Brown potatoes slowly, turning when necessary, adding more fat if needed. When potatoes are almost tender add thinly sliced onion, turning to heat until just wilted.

Add beef to heat through just before serving adding salt and pepper to taste. Makes 6 servings.

# Swiss Potato Soup

4 large white potatoes
3 cups water
1 teaspoon salt
¼ teaspoon marjoram leaves
2 tablespoons chopped celery leaves (only inner leaves)
5 green onions
2 tablespoons *each* butter and all-purpose flour
3 cups milk or half-and-half (light cream)
½ cup finely chopped parsley
¼ teaspoon pepper
  About 1½ cups shredded Swiss cheese

Peel potatoes and slice (about 5 cups). Place in a large saucepan and add water, salt, marjoram, celery leaves, and the white part only of 3 onions, finely chopped. Bring to a boil and simmer 25 minutes, or until very tender.

Remove from heat and mash with a potato masher.

Melt butter and blend in flour; cook 2 minutes, stirring. Gradually pour in milk and cook until thickened, stirring constantly. Add to potato mixture. Finely chop remaining 2 onions (using both white and green parts) and add to soup along with parsley and pepper. Cover and let stand a few minutes, then serve. Pass cheese at the table to spoon into each serving. Makes 8 servings of 1 cup each.

## Scalloped Potatoes with Cheese and Ham

6 large potatoes or about 4 pounds potatoes, peeled and
    cut into ¼-inch slices
2 teaspoons salt
1 medium-sized onion, thinly sliced
½ green pepper, seeded and diced
    Boiling water
4 tablespoons butter or margarine
½ cup all-purpose flour
3½ cups milk
1¼ teaspoons salt
⅛ teaspoon pepper
1½ cup diced leftover ham or 1 can (12 oz.) luncheon
    meat, cubed
1½ cups cubed sharp Cheddar cheese

Peel and cut enough potatoes to make 2 quarts. Add potatoes and the 2 teaspoons salt, onion, and green pepper to boiling water. As soon as water resumes boiling, remove from heat and drain vegetables. Turn into a large greased shallow casserole or baking pan (about 13 by 9 by 2 inches). Melt butter in a small pan, stir in flour until bubbly, then gradually stir in milk; cook, stirring constantly until thickened. Season with 1 1/4 teaspoons salt and the pepper, and pour over the potatoes in the casserole. Lightly mix in the diced ham or luncheon meat and about half the Cheddar cheese. Sprinkle remaining cheese over the top.

Bake, uncovered, in a 350° oven 35 to 40 minutes (or until potatoes are tender, the sauce is bubbly, and the cheese is melted). Makes 8 to 10 servings.

# SWEET

## Selection and Storage

**Selection:** Choose well shaped, firm sweet potatoes and yams that have smooth, bright, uniformly colored skins, free from decay.

**Buying:** Allow 1 medium-sized sweet potato (or 1/3 to 1/2 lb.) for each serving.

**Storage:** Keep in a dry, cool, area; do not refrigerate for they are damaged by chilling at temperatures below 50°.

## Cooking Suggestions

There are two types of sweet potatoes available. The one usually called *sweet potato* in markets is a dry mealy type that has the consistency of white Irish potatoes when cooked; it has pale colored flesh. Moist type potatoes are usually called *yams*; they have orange colored flesh and are sweet and moist after cooking. Some recipes specify one type or the other of sweet potatoes, but in other recipes you can use whichever you prefer.

**To bake sweet potatoes or yams,** scrub skins well and pat dry. Rub with butter or salad oil. Pierce skin with the tines of a fork and place in a shallow pan in a 400° oven until soft (about 30 to 40 minutes). Or bake them at a lower temperature along with the meat for about 1 hour.

**For boiled sweet potatoes or yams,** scrub and cook whole unpeeled potatoes in a generous amount of boiling salted water, covered, until tender (20 to 30 minutes). Or cook peeled potatoes, cut in quarters or sliced, in a small amount of boiling salted water until tender when pierced (10 to 20 minutes depending on size of pieces); drain.

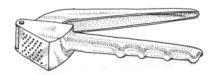

## Serving Suggestions

Sweet potatoes are used so often for casseroles that sometimes simpler cooking methods are overlooked. Here are some that are fast and effective.

**Buttered Sweet Potatoes**

Peel small sweet potatoes or yams (or larger ones, cut in quarters); cook following directions above; drain, and season to taste with salt, pepper, and melted butter.

**Fried Sweet Potatoes**

Peel cold, baked sweet potatoes and cut in crosswise slices about 3/4 inch thick. Melt enough butter or margarine in a frying pan to cover the bottom. Add potato slices and cook over medium heat, turning as little as possible, until browned lightly. Salt and pepper to taste.

**Baked Stuffed Yams**

Bake 4 medium-sized yams following directions above. Cool slightly, then cut slice off side and scoop out pulp, making shells. In a bowl beat the pulp with 2 tablespoons butter, 1/2 teaspoon salt, 1/4 teaspoon pepper, and 1 1/2 teaspoons grated orange peel. Then fold in 2 egg whites, beaten until stiff. Stuff back into shells and sprinkle tops with grated Parmesan cheese. Bake in a 350° oven for about 15 minutes.

### Oven Roasted Yams

Cut cooked, peeled yams into 2-inch-thick slices. About 1 1/2 hours before a turkey or other roast is done, arrange the yam slices around the meat in the roasting pan, turn slices over to coat with pan drippings and baste during cooking with pan drippings or meat basting sauce.

### Mashed Sweet Potatoes with Orange

Cook 5 medium-sized sweet potatoes as directed under cooking suggestions; peel and mash or beat

with an electric mixer. Beat in 1/4 cup (1/8 lb.) butter and 1/4 cup orange juice. Fold in 1 stiffly beaten egg white. Turn into a baking dish. Just before serving, bake at 350° for about 10 minutes, or until heated through.

### Candied Yams

Cook about 6 medium-sized yams following directions under *cooking suggestions*, page 79. Peel and cut into 1/4-inch-thick slices. Arrange slices overlapping in a shallow casserole. Dot the top with 4 tablespoons butter, sprinkle lightly with salt, then distribute 1/3 cup firmly packed brown sugar over top. If desired, drizzle with 3 tablespoons rum. Bake, uncovered, in a 375° oven for about 40 minutes, or until bubbly and glazed.

## Sweet Potato Salad with Curry Mayonnaise

    3 to 4 medium-sized sweet potatoes or yams
       Boiling water
    1½ cups celery, cut in slanting slices
     1 cup pineapple chunks, fresh or canned
    ½ cup whole or broken pecans or almonds
    ¼ cup orange juice
    ½ teaspoon salt
    2 teaspoons *each* curry powder, tarragon vinegar, and
       grated orange peel
    1 cup mayonnaise
    1 or 2 tablespoons milk or half-and-half (light cream)
       Lettuce leaves
       Major Grey's Chutney

Cook sweet potatoes in water to cover for 20 to 30 minutes or just until tender when pierced; do not overcook. Peel slightly cooled potatoes and cut into large cubes enough to make about 3 cups. Mix gently with celery, pineapple chunks, nuts, orange juice, and salt. Chill, covered, until ready to serve.

To prepare mayonnaise, blend curry powder, vinegar, orange peel, and mayonnaise. Thin dressing with milk.

To serve, mound salad onto lettuce leaves. Top each portion with curry mayonnaise; pass chutney to accompany the salad. Makes 6 generous servings.

## Cashew-Yam Casserole

    About 2½ pounds yams or sweet potatoes, whole or
       halved
       Boiling salted water
    1 teaspoon ground cinnamon
    ¼ teaspoon salt
    1 egg
       About ¼ cup pineapple juice or orange juice
       About ¼ cup sugar
    3 tablespoons melted butter or margarine
    ½ cup salted cashews, coarsely chopped

In a saucepan, cook the potatoes in boiling water until very tender (about 30 minutes); drain. When cool enough to handle, peel potatoes. Using an electric mixer or potato masher, beat until mashed; then measure 3 cups. Add cinnamon, salt, egg, juice, and sugar. Beat

until fluffy, adding more fruit juice if mixture seems dry. Taste and add more sugar or salt, if needed. Mix in 2 tablespoons of the butter.

Spoon into a 1-quart casserole or soufflé dish. (Cover and refrigerate at this point, if desired). Add the cashews to remaining 1 tablespoon butter in a small frying pan; heat, stirring, until lightly toasted. Sprinkle on top of casserole. Bake, uncovered, in a 375° oven for about 15 minutes until heated through (about 35 minutes if refrigerated). Makes about 6 servings.

## Yam Soufflé

    3 cups mashed yams or sweet potatoes (unseasoned)
    ⅓ cup Sherry or orange juice
    1¼ cups half-and-half (light cream)
    6 tablespoons melted butter or margarine
    1½ teaspoons grated orange peel
    ⅛ teaspoon *each* pepper and ground nutmeg
    1 teaspoon salt
    2 tablespoons firmly packed brown sugar
    6 eggs, separated

Combine in the mixer bowl the mashed yams, Sherry, cream, butter, orange peel, pepper, nutmeg, salt, and sugar; beat until blended and smooth. Add the egg yolks and beat until fluffy. This much can be done ahead and the mixture held at room temperature for several hours. Just before baking whip egg whites until they hold short, distinct, moist peaks; carefully fold into potato mixture.

Spoon into about 12 well buttered, individual soufflé dishes (about 5 oz. size) and bake in a 375° oven for about 20 minutes. (Or bake in a 2 1/2-quart soufflé dish about 45 minutes.) Serve immediately. Serves 12.

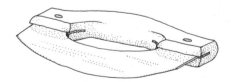

# RUTABAGA

## Selection and Storage

**Selection:** Choose large rutabagas that are heavy for their size, firm, and have generally smooth round or elongated shape. Avoid rutabagas with skin punctures or deep cuts and decay.

**Buying:** Allow 1/2 pound for each serving.

**Storage:** Store roots in cool, moist area or put in plastic bag in refrigerator.

## Cutting and Cooking

Wash and remove thin layer of skin with vegetable peeler. Rutabaga can be left whole, quartered, sliced, diced, cut in strips, or shredded.

**To cook rutabagas,** place in boiling salted water. Cook, covered, until tender (about 30 minutes for whole, 7 to 20 minutes for cut pieces, depending on size); drain.

## Selection and Storage

The yellow-fleshed rutabaga can be cooked in most of the same ways as turnips. Rutabagas are sweeter, crisper, and more flavorful. Season this cooked vegetable simply with salt, pepper, butter; add a squeeze of lemon juice, if you like. Other ideas follow.

### Baked Rutabaga

Place 3 cups rutabagas, cut in 1/4-inch-thick sticks in a shallow, buttered casserole. Dot with 3 tablespoons butter; sprinkle with 2 tablespoons water and salt and pepper. Cover tightly. Bake in a 400° oven for about 25 minutes or until rutabagas pierce easily with a knife. Makes 5 to 6 servings.

### Shredded Rutabagas

In a wide heavy pan, combine 2 cups coarsely shredded, firmly packed rutabagas, 2 tablespoons water, 3 tablespoons melted butter, 1 to 1 1/2 tablespoons brown sugar, and 1 teaspoon soy sauce. Cover, cook quickly, stirring frequently until tender-crisp (about 7 minutes). Serves 4 to 5.

### Sautéed Rutabagas

In a saucepan, combine 2 cups rutabagas cut in 1/2-inch cubes and 1/3 cup boiling salted water; cover and cook until tender (about 15 minutes); drain. Heat 2 tablespoons butter or margarine in a frying pan and sauté rutabagas until lightly browned. Sprinkle with 1/2 teaspoon poppy seed or 1/4 teaspoon anise seed. (To cream rutabagas, add 1/4 cup whipping cream and simmer rapidly, stirring until sauce thickens enough to coat vegetables lightly.) Makes 4 servings.

### Rutabagas in Broth

Cover and cook 2 cups thinly sliced rutabagas in 1/2 cup regular strength beef broth until tender (about 15 minutes). Add 1 small bay leaf to broth, if desired, removing it before serving. Makes 4 servings.

### Glazed Rutabagas

Pare and slice or dice 2 pounds rutabagas. Par-boil in a quantity of salted water until almost tender; drain, saving 1/2 of the cooking liquid. Add to rutabagas 1/4 cup butter or margarine, 1/2 teaspoon sugar, 1/2 teaspoon salt, and the 1/2 cup liquid. Cook, shaking pan occasionally, until rutabagas are glazed and liquid has evaporated.

## Rutabagas with Lemon Dill Butter

4 medium-sized rutabagas, peeled
1/3 cup melted butter or margarine
1 teaspoon lemon juice
1/4 teaspoon dill weed

Cut rutabagas in thin slices and cook following *cutting and cooking* above; drain. Mix together butter, lemon juice, and dill weed; pour over rutabagas. Serves 4.

## Mashed Rutabaga and Potatoes

1 large rutabaga, peeled and diced
4 medium-sized potatoes
  Butter or margarine
  Salt and pepper to taste
  Hot milk, if needed
  Chopped parsley

Cook rutabaga following directions under *cutting and cooking* above; drain. Boil potatoes in their jackets for about 30 minutes until tender; drain and slip off skins.

Combine hot rutabaga and hot potatoes. Mash or beat together with butter, salt, and pepper to taste. If the mixture is dry, beat in a little hot milk to obtain consistency you desire. Serve sprinkled with chopped parsley. Makes 6 to 8 servings.

## Fluffy Rutabaga Bake

About 2 pounds rutabagas
Boiling salted water
2 tablespoons all-purpose flour
1 teaspoon seasoned salt
¼ teaspoon ground cinnamon
Dash pepper
2 eggs
¼ cup whipping cream

Peel and cut rutabaga in large chunks. Cook, following directions under *cutting and cooking*, page 81; drain. Beat with an electric mixer or mash with a fork until smooth, (you should have about 3 cups purée).

Turn vegetable purée back into the mixer bowl, add flour, salt, cinnamon, and pepper, and beat until smooth. Add the eggs one at a time and beat well. Add cream and continue beating until light and fluffy.

Spoon into a greased 1-quart casserole.

Bake, uncovered, in a 350° oven for about 30 minutes until heated through. Serve immediately. Makes 6 to 8 servings.

## Rutabagas with Soy-Butter Sauce

2 medium-sized rutabagas (about 2 lbs.)
Soft butter or margarine
¼ cup (⅛ lb.) butter or margarine
2 teaspoons firmly packed brown sugar
2 tablespoons soy sauce
1 tablespoon lemon juice
1 teaspoon Worcestershire

Scrub rutabagas, peel, cut in half, and lightly spread with soft butter. Wrap each half tightly in foil. Bake in a 400° oven for 1 hour or until tender when pierced. Meanwhile, melt the 1/4 cup butter in a pan; add brown sugar and stir until well blended. Stir in soy, lemon juice, and Worcestershire. Simmer, stirring occasionally, for 3 minutes. Just before serving, unwrap rutabagas and cut into chunks. Pour the soy-butter sauce over all. Makes 4 servings.

# SQUASH

*Summer: Crookneck, Patty Pan, Zucchini*
*Winter: Banana, Butternut, Hubbard, Pumpkin*

# SUMMER

### Selection and Storage

**Selection:** Choose squash with fresh, glossy, smooth skin that is not hard or tough. The squash should be heavy for its size.
**Buying:** One pound provides 3 to 4 servings.
**Storage:** Place in a plastic bag and refrigerate. Will keep for several days.

### Cutting and Cooking

Remove stem and blossom ends and scrub squash with a vegetable brush just before cooking. Do not remove the tender peel. Leave whole, cut into slanting slices, circles, cubes, or slice lengthwise.
**To cook whole,** place in a saucepan of boiling salted water to cover. Simmer about 10 to 12 minutes or until tender-crisp; drain.
**To par-boil squash** before baking with a stuffing, cook whole squash until almost tender (about 7 to 10 minutes); drain.

**To cook slices,** place in a small amount of boiling salted water; cook, covered, until tender-crisp (about 3 to 8 minutes depending on size of the pieces); drain.

### Seasoning and Serving

Season cooked squash with salt, pepper, butter or margarine, or grated Parmesan cheese. Many spices and herbs complement this vegetable; basil leaves, chives, chile powder, curry powder, dill weed, dry mustard, marjoram leaves, ground nutmeg, oregano leaves, thyme leaves.

**Mashed Summer Squash**

Cook 1 1/2 pounds diced crookneck or patty pan squash following directions under *cutting and cooking*, above, for slices. Drain, if necessary; mash coarsely and season to taste with salt, pepper, and melted butter. Makes 4 to 5 servings.

**Hot Chile Zucchini**

Cut 1 1/2 pounds zucchini in 1/2-inch slices and cook in a large quantity of boiling salted water until almost tender (2 to 4 minutes); drain and cool quickly in cold water if done ahead.

At serving time, melt 3 tablespoons butter or margarine in a 10-inch frying pan and add 3 tablespoons *each* chopped onion and green pepper. Cook, stirring, over medium heat until onion becomes translucent. Turn heat high, add zucchini, and cook, stirring, until hot through (about 3 minutes). Add 2 to 3 tablespoons finely chopped California green chiles (seeds and pith removed) and salt to taste. Makes 4 to 6 servings.

### Zucchini Fritters

Trim stem and blossom end from medium-sized zucchini (1 per serving) and cut in 1/4-inch-thick lengthwise slices. Dust strips lightly in all-purpose flour, dip in beaten egg (mix each egg with 1 1/2 tablespoons water). Fry over medium heat in a wide frying pan filled with 1 inch hot salad oil. Turn fritters to brown evenly; drain and salt. (Keep covered if made ahead.) Reheat fritters in a single layer on baking sheets lined with absorbent material in a 350° oven for 5 minutes. Serve with lemon wedges.

### French Fried Crookneck Squash

Cut 2 pounds crookneck squash lengthwise in 1/4-inch-thick slices. Dip slices in 2 eggs beaten with 1/2 teaspoon salt and 2 tablespoons water; roll in about 3 cups finely crushed cracker crumbs. Keep slices well separated and let dry about 30 minutes. Fry until nicely browned in deep fat heated to 370°; drain. Serve hot. Makes 6 servings.

### Skillet Zucchini

In a large frying pan sauté 1 clove garlic, minced or mashed, in 3 tablespoons olive oil for about 3 minutes. Add 6 medium-sized zucchini or other summer squash, sliced about 1/4 inch thick, and cook, uncovered, over medium heat, stirring occasionally, 8 to 10 minutes. Stir in a mixture of 1 tablespoon finely chopped parsley, 1 teaspoon crumbled oregano leaves, 3/4 teaspoon salt, 1/4 teaspoon sugar, and 1/8 teaspoon seasoned pepper; cook for 3 to 5 minutes longer or until zucchini is just tender. Makes 4 to 6 servings.

### Crookneck and Zucchini à la Grecque

Cut off ends of 6 crookneck squash and 6 zucchini. Slice in half lengthwise and cook, following directions under *cutting and cooking* slices, page 82. Blend together 4 tablespoons olive oil, 2 tablespoons lemon juice, 1/2 teaspoon crumbled oregano leaves, and a dash salt and pepper. Pour dressing over hot squash. Makes 6 to 8 servings.

### Creamed Patty Pan

Trim ends of 2 pounds fresh patty pan or other summer squash. Scrub and shred coarsely using the largest holes of your grater (you should have about 8 cups). Using a 10-inch or larger frying pan with a tight fitting lid, combine 4 tablespoons (1/8 lb.) butter or margarine, 2 tablespoons water, 1/8 teaspoon pepper, 1/2 teaspoon *each* salt and basil leaves, and 1 small clove garlic, minced or mashed. Place on high heat; mix in the squash, cover, and cook until the squash is just tender (about 5 minutes). Take cover off the last minute or two, if necessary, to evaporate most of the liquid.

Meanwhile, mix 1 cup sour cream with 1 tablespoon flour until smooth; stir into squash. Bring to a boil, stirring until smooth. Serves 4 to 6.

---

### Butter-Steamed Zucchini

Melt 2 tablespoons butter or margarine in a wide frying pan or electric frying pan over high heat. Add 4 cups thinly sliced zucchini, stem ends removed (about 8 zucchini), and 6 tablespoons water. Cook covered, stirring occasionally, for 5 minutes. Season with salt. Makes about 5 servings.

### Butter-Steamed Zucchini with Cheese

Prepare zucchini as for Butter-Steamed Zucchini. After cooking 5 minutes, remove cover and crumble in 2 tablespoons blue cheese or add 4 tablespoons shredded Parmesan cheese; stir until cheese is melted. Season with salt and serve. Makes about 5 servings.

# Summer Squash with Dill

1½ pounds summer squash
1 tablespoon butter or margarine
⅓ cup finely chopped onion
1 tablespoon water
1 teaspoon salt
¼ teaspoon pepper
1 cup (½ pt.) sour cream
1 teaspoon dill weed
1 tablespoon all-purpose flour
Chopped parsley

Wash squash and trim off the ends but do not peel; cut into 1/2-inch slices. Put the butter in the bottom of a 1 1/2-quart casserole and place in the oven while it is preheating to 350°. When butter is melted, remove it from the oven and arrange the sliced squash and onion in the casserole. Sprinkle over the water, salt, and pepper. Cover the casserole and bake for about 40 minutes.

In a bowl, mix the sour cream, dill weed, and flour until well blended. Remove the casserole from the oven and stir the sour cream mixture into squash thoroughly until well blended.

Return to the oven and bake, uncovered, for about 15 minutes more or until squash is tender but still slightly crisp. Garnish with chopped parsley before serving. Makes about 6 servings.

## Almond-Topped Squash

3 tablespoons butter or margarine
⅓ cup slivered blanched almonds
6 medium-sized patty pan, cut into large chunks, or crookneck, cut into thick slices or zucchini cut lengthwise into sticks
2 tablespoons olive oil
1 small onion, finely chopped
¼ teaspoon salt
⅛ teaspoon pepper

In a large frying pan, melt 1 tablespoon of the butter over medium heat. Add the almonds and cook, stirring frequently until nuts are golden brown (about 6 minutes). Lift from pan with a slotted spoon and set aside.

Add remaining butter and olive oil to frying pan; add squash, onion, salt, and pepper; cook over medium-high heat, turning frequently until squash is tender-crisp when pierced with a fork but not brown (about 12 to 14 minutes). Spoon squash and onion into a serving dish; sprinkle with toasted almonds. Serves 6.

## Stuffed Patty Pan Squash

2 tablespoons butter or margarine
2 pounds patty pan squash (about 3 inches in diameter)
1 package (3 oz.) cream cheese with chives, softened
½ teaspoon Worcestershire
1 tablespoon half-and-half (light cream)
3 tablespoons chopped almonds
½ teaspoon salt
Dash liquid hot pepper seasoning
¼ cup grated Parmesan cheese

Put butter into a 9 by 13-inch baking pan; set into a 400° oven until melted. Meanwhile, wash the squash, trim ends, and cut each in half lengthwise, keeping the halves together. In a small bowl, combine the cream cheese, Worcestershire, cream, almonds, salt, and hot pepper seasoning; beat until well blended. Spread between two squash halves, sandwich fashion. Turn squash over in the melted butter to coat both sides. Cover the pan with foil or a lid.

Bake in a 400° oven for about 1 hour or until tender. Remove from oven, sprinkle with Parmesan, and put back into the oven, uncovered, until lightly browned (about 5 minutes). Makes about 6 servings.

## Zucchini Layered Casserole

¾ cup soft bread crumbs
2 medium-sized zucchini, sliced
1 large onion, thinly sliced
2 medium-sized tomatoes, peeled and sliced
½ teaspoon salt
¼ teaspoon *each* pepper and oregano leaves
2 tablespoons butter or margarine

Place 1/2 cup of the soft bread crumbs in the bottom of a well greased 1 1/2-quart casserole. Arrange zucchini, onion, and tomatoes in layers, using half of each vegetable per layer. Sprinkle each layer with salt, pepper, and oregano. Repeat with remaining vegetables; sprinkle remaining 1/4 cup soft bread crumbs over the top and dot with the butter. Bake, uncovered, in a 350° oven for 1 hour or until vegetables are tender. Makes 4 to 6 servings.

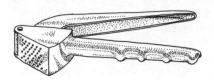

## Summer Squash Gratinée

½ cup water
1 pound patty pan squash, thinly sliced
¼ cup whipping cream
½ teaspoon salt
1 cup (¼ lb.) shredded Cheddar cheese

Cook squash following directions under *cutting and cooking*, page 82. Arrange squash in an ovenproof casserole (at least 1-qt. size). Whip cream with salt until it holds firm peaks; spread evenly over surface of squash; sprinkle cheese over all. Place about 2 inches from broiler until cheese is browned (about 2 minutes). Serve immediately. Makes 4 servings.

## Shredded Zucchini Salad

1 fully ripe avocado, peeled
Lemon juice
4 cups shredded unpeeled zucchini (about 4 zucchini)
Mustard Mayonnaise (recipe follows)
1 cup cherry tomatoes, halved

Carefully slice the avocado into crescents; turn each slice in lemon juice and set them aside. In a salad bowl, mix together the zucchini and Mustard Mayonnaise; then add tomatoes and toss lightly. Arrange avocado on top. Makes 4 servings.

**Mustard Mayonnaise.** In a small bowl, stir together until thoroughly combined 1/3 cup mayonnaise, 1 teaspoon prepared Dijon-style mustard, 1/8 teaspoon pepper, and 1/4 teaspoon Worcestershire.

## Calabazas Rellenas (Stuffed Squash)

6 medium-sized zucchini
1 package (3 oz.) cream cheese
2 tablespoons minced onion
½ teaspoon salt
¼ teaspoon pepper
1 cup (½ pint) sour cream
Paprika (optional)

Cook whole zucchini following directions under *cutting and cooking*, page 82; drain. Allow to cool until you are able to handle them, then cut each zucchini in half lengthwise and scoop out seeds into a small bowl. Mix the seeds with the cream cheese, onion, salt, and pepper. Stuff this mixture back into the zucchini halves; arrange them in a buttered square baking dish or pan. Spoon the sour cream evenly over the top of each. Sprinkle with paprika, if desired. Bake for about 10 minutes in a 325° oven. Serve immediately. Allow 2 zucchini halves for each serving. Makes 6 servings.

## Patty Pan Squash and Corn

8 medium-sized patty pan squash
1 can (12 oz.) whole kernel corn
¾ cup milk
1 tablespoon finely minced onion
1 teaspoon salt
   Pepper to taste
1½ cups diced jack cheese

Slice squash and put in a saucepan with the corn, milk, onion, salt, and pepper; cover and simmer slowly until squash is tender (about 10 minutes). Add cheese and heat just until the cheese is melted. Serves 6 to 8.

## Scalloped Zucchini

1 medium-sized carrot, thinly sliced
2 tablespoons *each* chopped onion and green pepper
½ cup sliced celery
¼ cup salted water
3 medium-sized zucchini
1 large tomato, peeled
3 tablespoons butter or margarine
3 tablespoons all-purpose flour
1½ cups milk
   Salt and pepper to taste
1 cup crushed corn chips

In a saucepan simmer the carrot, onion, green pepper, and celery in the 1/4 cup salted water until vegetables are tender and water has evaporated. Turn into a greased 1 1/2-quart casserole. Thinly slice zucchini, arrange over carrot mixture, then cover with sliced tomato. Melt butter, blend in flour, and gradually stir in milk; cook until thickened. Season to taste with salt and pepper. Pour cream sauce over the vegetables. Sprinkle crushed corn chips over the top. Bake in a 350° oven for 45 minutes or until zucchini is tender. Makes 6 to 8 servings.

## Zucchini-Spinach Bake

3 tablespoons butter or margarine
2 packages (10 oz. *each*) frozen chopped spinach,
   completely thawed
⅛ teaspoon ground nutmeg
6 cups zucchini, sliced ⅛ inch thick (about 5 zucchini)
2 tablespoons water
¼ teaspoon salt
1 cup grated fresh Parmesan cheese

In a large frying pan with a lid, melt 2 tablespoons of the butter over high heat. Add thawed spinach and sprinkle on nutmeg. Heat quickly, stirring until liquid boils away. Remove spinach from pan and set it aside.

To same frying pan, add remaining 1 tablespoon butter and melt over high heat. Add zucchini, water, and salt; stir and cover. Cook at high heat for about 3 minutes, stirring occasionally until zucchini just begins to become limp and slightly transparent; remove from heat.

Layer half the zucchini in a buttered 8-inch-square baking pan. Add spinach and pat out to an even layer; sprinkle on half the cheese. Then spoon on remaining zucchini in an even layer; sprinkle on remaining cheese. Cover pan with foil and bake in a 350° oven for 15

minutes. Remove foil and continue baking for 10 minutes. (Or cover and refrigerate; then bake covered for 20 minutes; remove foil and bake for an additional 10 minutes.) Makes 8 servings.

## Baked Zucchini with Mushrooms

2 tablespoons butter or margarine
½ pound mushrooms, sliced
½ teaspoon salt
⅛ teaspoon *each* garlic salt and pepper
¼ teaspoon Italian herb seasoning or oregano leaves
4 medium-sized zucchini (about 1 lb.)
¼ cup fresh bread crumbs (whirl bread pieces in blender)
4 tablespoons grated Parmesan or Romano cheese
4 eggs, slightly beaten

In a frying pan, heat the butter; add the mushrooms and sauté over medium-high heat for about 5 minutes until most of the liquid has evaporated. Stir in the salt, garlic salt, pepper, and Italian herb seasoning; set aside.

Wash and trim zucchini, then finely shred it. In a bowl, combine zucchini, bread crumbs, 2 tablespoons of the cheese, and mushroom mixture. Spoon into a greased 8-inch-square baking dish. (If making the dish ahead, cover and let stand at room temperature.)

Pour eggs over zucchini and stir gently with fork. Bake, uncovered, in a 325° oven just until custard is set (35 to 40 minutes). Serve topped with remaining cheese. Makes 6 to 8 servings.

## Crusty Zucchini Strips

3 medium-sized zucchini
   About ⅓ cup all-purpose flour
1 egg beaten with 1 tablespoon milk
   About ½ cup corn flake crumbs
   About 4 tablespoons olive oil or salad oil
½ cup plain yogurt (optional)

Cut each zucchini lengthwise into eighths. Place flour in one flat pan, egg-milk mixture into a second, and crumbs in a third. Roll zucchini, about 8 strips at a time, in the flour and shake off excess; then in the egg-milk mixture; and finally in the corn flake crumbs, coating evenly; lay on waxed paper.

Heat 3 tablespoons olive oil in a wide frying pan over medium-high heat; add zucchini (about 8 at a time) and cook 1 to 2 minutes until tender when pierced, turning as needed to brown evenly; add more oil as needed. Drain and keep warm in a pan, uncovered, in a 200° oven while you cook remainder. Serve with yogurt, if desired. Makes 4 servings.

# Zucchini-Rice Casserole

Water
½ teaspoon salt
½ cup long grain rice
1 medium-sized onion, chopped
1 pound zucchini (3 to 4 medium)
½ green pepper, diced
1 can (8 oz.) tomato sauce
⅔ cup shredded cheese (such as Cheddar, Swiss, jack)
¼ cup grated or shredded Parmesan cheese (optional)

In a saucepan, bring to boiling 1 cup water with 1/4 teaspoon salt. Add the rice and onion; reduce heat, cover, and simmer until rice is tender (about 20 minutes). Meanwhile, scrub zucchini and cut into 1/4-inch slices. In another pan, bring 1/2 cup water to boiling with 1/4 teaspoon salt. Add zucchini and green pepper; cook until zucchini is almost tender (about 5 minutes), stirring occasionally. Remove from heat and immediately pour into colander to drain.

When rice is tender and its liquid absorbed, add tomato sauce and zucchini; lightly mix together and turn into a 1 1/2-quart casserole. Sprinkle top with the cheeses. Bake, uncovered, in a 325° oven until heated through (about 20 minutes). Makes about 6 servings.

# Winter

## Selection and Storage

**Selection:** Choose heavy winter squash (Acorn, Butternut, Banana, Hubbard, Pumpkin) with hard, thick rind. Pulp should be thick and bright yellow-orange in color.

**Buying:** Allow 1 1/2 pounds for 4 servings.

**Storage:** Keep whole squash in a cool dry area. If cut in pieces, wrap in a plastic bag or film and keep refrigerated. Will keep for several days.

## Cutting and Cooking

Cut large squash such as Hubbard or Banana apart with a heavy-bladed knife or hand saw; then cut into serving size pieces. Scrape away the seeds and stringy portions. Smaller squash (such as Butternut or Acorn), cut in half, usually make serving size pieces. Wash and cut just before serving.

**For baked squash,** arrange pieces with skin side down in a greased baking dish. Score the flesh with a knife and sprinkle with salt and pepper, if desired. Spread surface with butter and bake in a 375° to 400° oven until tender (45 minutes to 1 hour).

**To bake squash more quickly,** place unseasoned pieces cut side down in baking pan or place in a covered baking dish and bake in a 400° to 450° oven until tender (about 30 to 40 minutes).

**To cook squash in a pan,** remove the peel from cut pieces with a vegetable peeler or knife. Dice or slice squash. Cook in a small amount of boiling salted water until just tender (about 4 to 5 minutes, depending on size of pieces); drain.

**To steam squash,** place peeled or unpeeled pieces in a steamer over boiling water; cook until tender (about 10 minutes for 1-inch cubes).

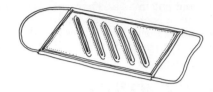

## Seasoning and Serving

Baked squash can be served in the shell with butter, or you can season the cut pieces as they bake with butter and other flavorings, such as ground coriander, allspice, or nutmeg, and with sweeteners like chopped candied ginger, brown sugar, or maple syrup. Unseasoned baked squash can be scooped out of the shell, mashed, then seasoned.

A delicious and quick way to prepare squash is to peel and dice it, then cook it for a few minutes in a frying pan with a little butter and water. Here are some of the ways you can vary the flavor of squash when you butter-steam it.

### Butter-Steamed Squash

Peel and cut about 1 1/2 pounds winter squash into 1/2-inch cubes to make 4 cups. Melt 2 tablespoons butter or margarine in a wide frying pan over medium-high heat. Add the squash and 1/4 cup water. Cover and cook, stirring occasionally, 4 to 5 minutes or until just tender and liquid is absorbed. Stir in salt to taste. Makes 4 servings.

### Cream-Glazed Cardamom Squash

Follow recipe for Butter-Steamed Squash. After the squash is tender, stir in salt to taste, 1/8 teaspoon ground cardamom, and 1/4 cup whipping cream; cook, uncovered, stirring until liquid is almost gone and squash is glazed.

### Cream-Glazed Anise Squash

Follow recipe for Butter-Steamed Squash, adding 1/4 teaspoon anise seed with the squash and water. After squash is tender, add 1/4 cup whipping cream with the salt; cook, uncovered, stirring until liquid is almost gone and squash is glazed.

### Squash with Bacon and Onion

Dice 6 slices bacon and fry in a wide frying pan until crisp. With a slotted spoon remove bacon from pan and drain, reserve only 4 tablespoons drippings in the pan. Add 1 small onion (chopped) and 1 clove garlic (minced or mashed); cook over medium heat until limp. Increase heat to medium-high and add to bacon drippings 4 cups (1 1/2 lbs.) winter squash (peeled and cut in 1/2 inch cubes), and 1/4 cup water. Cover and cook, stirring occasionally, until tender (4 to 5 minutes). Stir in bacon, 1 teaspoon vinegar, and salt to taste. Makes about 4 servings.

### Squash with Apples

Follow recipe for Butter-Steamed Squash, page 86. Use 1 pound squash and increase butter to 4 tablespoons. With the squash and water add 1 large Golden Delicious apple (peeled and cut in 1/2-inch cubes), and 2 teaspoons lemon juice. After squash is tender, stir in 1/4 cup firmly packed brown sugar, 1/4 teaspoon ground cinnamon, 1/8 teaspoon ground nutmeg, and salt to taste. Garnish with 1/4 cup toasted sliced almonds sprinkled over the top.

### Squash with Peanuts and Coconut

Follow recipe for Butter-Steamed Squash, page 86. Instead of salt, add 1 teaspoon *each* soy sauce, sugar, and vinegar; stir to blend. Garnish with 2 tablespoons *each* chopped roasted salted peanuts and toasted coconut.

# Cream of Pumpkin Soup

2 tablespoons butter or margarine
¼ cup finely chopped onion
1 teaspoon curry powder
1 tablespoon all-purpose flour
2 cans (about 14 oz. *each*) regular strength chicken broth
1 can (1 lb.) pumpkin
1 teaspoon brown sugar
⅛ teaspoon ground mace or nutmeg
¼ teaspoon salt
⅛ teaspoon pepper
1 cup milk or half-and-half (light cream)
Minced chives or parsley

Heat butter in a 2-quart or larger pan and sauté onion over medium heat until limp. Stir in curry and flour and cook until bubbly. Remove from heat and gradually stir in the chicken broth. Add pumpkin, sugar, mace, salt, and pepper. Cook, stirring, until the mixture begins to simmer. Stir in the milk and continue heating, but do not boil.

Sprinkle a few minced chives or parsley into each bowl when you serve. Makes about 6 servings.

# Glazed Banana Squash

1½ pounds banana squash
4 tablespoons (⅛ lb.) butter or margarine
About 4 tablespoons water
1 tablespoon frozen orange juice concentrate, thawed
3 tablespoons apricot jam
⅛ teaspoon ground cloves
¼ teaspoon salt
Dash pepper

Using a vegetable peeler or knife, remove peel from the squash. Cut squash into 1/4-inch cubes (you should have about 4 cups). Melt butter in an electric frying pan set at highest heat (or use a wide frying pan at least 10 inches in diameter, over high heat). Add squash and 2 tablespoons of the water. Stir, cover, and cook over high heat for about 4 to 5 minutes or until just tender and liquid is absorbed. Stir several times, adding more water if necessary to keep squash from sticking.

Reduce the heat to lowest setting and stir in the orange juice concentrate, jam, cloves, salt, and pepper. Stir gently until squash is evenly glazed. Makes 4 to 6 servings.

# Pumpkin Succotash

4 slices bacon, chopped
½ cup chopped onion
1 clove garlic, finely chopped
1 green pepper, seeded and finely chopped
3 cups peeled, seeded pumpkin, cut in ¾-inch cubes
1 can (1 lb.) tomatoes
1 cup uncooked, sliced green beans
1 package (10 oz.) frozen whole kernel corn, thawed
2 teaspoons salt
Dash pepper

Fry bacon until it is crisp; drain and reserve. Measure drippings and return 1/4 cup to frying pan. Add onion, garlic, green pepper, and pumpkin; cook, stirring occasionally, for 5 minutes. Stir in tomatoes, beans, corn, and salt and pepper. Cover and simmer until pumpkin is tender (about 25 minutes). Serve topped with bacon. Makes 6 to 8 servings.

# Pear-Filled Squash

¼ cup butter
2 medium-sized onions, sliced
2 pears, peeled and diced
½ teaspoon salt
¼ teaspoon *each* ground ginger and ground cinnamon
2 tablespoons *each* firmly packed brown sugar and Sherry or apple juice
3 small acorn squash, cut in halves and seeded
¼ cup toasted, sliced almonds

Melt butter in a frying pan over low heat. Add sliced onions and cook, stirring as needed, until golden (about 30 minutes). Stir in pears, salt, ginger, cinnamon, brown sugar, and Sherry or apple juice. Cook 2 more minutes; remove from heat.

Arrange acorn squash, cut side down, in a shallow baking pan. Pour in hot water to 1/4 inch deep. Bake in a 450° oven for 25 minutes or until tender when pierced.

Turn cooked squash over and mound pear mixture in centers; bake 10 minutes more. Top with toasted, sliced almonds just before serving. Makes 6 servings.

# TOMATOES

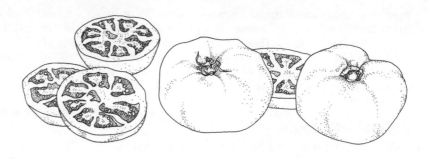

## Selection and Storage

**Selection:** Choose firm, well-formed tomatoes that are free from deep blemishes and not over-ripe.

**Buying:** Allow 1 tomato for each serving or 1/2 pound.

**Storage:** Keep in a cool place around 50°. Otherwise purchase them close to serving time and, if necessary, refrigerate briefly.

## Cutting and Cooking

Although usually considered a salad ingredient, tomatoes also make a fine cooked vegetable.

**To peel tomatoes,** dip into boiling water for about 1 minute, then plunge in cold water; slip off skin. Or hold tomato with a fork over a flame until the skin splits; peel.

**When recipes call for seeded tomatoes,** an easy way is to slice the tomatoes in half, then gently squeeze and press out the seedy portions. Then dice or cut in sections as called for. Other recipes simply call for peeled or unpeeled tomato slices or wedges.

## Serving Tomatoes Hot

A wide variety of seasonings enhance the flavor of cooked tomatoes. Choices include basil leaves, chives, dill weed, garlic, onion, oregano leaves, parsley, tarragon leaves, and Parmesan cheese. A dash of sugar may be added to balance the natural acidity of tomatoes.

### Buttered Tomatoes

Peel 2 pounds large, firm tomatoes. Gently press out and discard pulp and seeds; cut flesh in squares. Put in a heavy saucepan with 1/4 cup (1/8 lb.) butter or margarine, salt and pepper to taste, and a pinch sugar. Cover and cook just long enough for tomatoes to heat.

### Cherry Tomatoes in Butter

Cut 1 1/2 cups cherry tomatoes in halves; heat quickly in 2 teaspoons butter, uncovered, shaking pan until hot through. Season with salt and pepper to taste. Makes 2 servings.

### Creamy Sautéed Tomatoes

Place 3 cups peeled and coarsely chopped tomatoes (about 3 large tomatoes) in a frying pan or sauce-pan with 1/4 cup (1/8 lb.) butter or margarine, 3 tablespoons chopped onion, and 1/8 teaspoon garlic powder. Cook over medium heat, stirring occasionally, until tomatoes are tender but still brightly colored. Add salt, pepper, and ground nutmeg to taste. Blend in 1/3 cup whipping cream. Serve in bowls. Makes 4 servings.

### Crumb-Coated Fried Tomatoes

Cut 6 firm tomatoes in 1/2-inch slices. In one bowl put 1 cup milk. In a second bowl mix 1 cup all-purpose flour, 1 teaspoon *each* salt and sugar, and 1/4 teaspoon pepper. In a third bowl beat 2 eggs, in a fourth put 2 cups finely crushed salted soda crackers. Dip both sides of *each* tomato slice first in the milk, then in the flour, the beaten egg, and the cracker crumbs. Melt enough butter in a frying pan to coat bottom generously; fry the tomatoes about 2 minutes on each side over medium heat or until golden on both sides. Serves 6.

### Sour Cream Baked Tomatoes

Cut 4 or 5 medium-sized tomatoes, peeled and cored, in half crosswise; arrange in a shallow baking dish. Sprinkle with seasoned salt. Spread with mixture of 1/2 cup sour cream, 1/3 cup mayonnaise, a pinch of ground nutmeg, and 1/8 teaspoon of dill weed. Bake in a 375° oven for 15 to 20 minutes or until tender. Makes 4 to 5 servings.

### Herb Baked Tomatoes

Choose 6 to 8 tomatoes, each about 2 inches in diameter, and cut out stem end decoratively. Blend 5 tablespoons fine, dry bread crumbs with 2 tablespoons melted butter and 1/2 teaspoon dill weed. Spoon equal amounts of the crumb mixture into cut surface of each tomato and pat firmly in place. Chill, covered, until ready to bake. Bake in a 450° oven for about 20 minutes or until hot throughout. Sprinkle with salt. Serves 6 to 8.

### Broiled Tomatoes Parmesan

Halve 6 medium-sized tomatoes crosswise; sprinkle with salt and pepper. Then sprinkle about

1/4 teaspoon grated Parmesan cheese over each tomato half; dot each with 1/4 teaspoon butter. Broil about 6 inches from broiler for 4 to 5 minutes. Makes 6 servings of 2 halves each.

### Serving Sliced Fresh Tomatoes

The large, meaty tomatoes of summer are especially delicious cut in thin slices and seasoned lightly with a marinade or dressing. These four are good dressings for sliced tomatoes.

#### Basil Dressing

Combine in a blender container 1/4 cup olive oil or salad oil (or half of *each*), 2 tablespoons garlic (or regular) red wine vinegar, 1/2 teaspoon *each* salt and sugar, and 1/4 cup fresh basil leaves (loosely packed in cup) or 1 tablespoon dried basil leaves; whirl until blended. Let stand 1 to 2 hours to mellow flavors. Drizzle over thinly sliced tomatoes shortly before serving. Grind fresh pepper over top, if desired. Makes 4 to 6 servings.

#### Tarragon Dressing

Prepare dressing the same as for Basil Dressing above, omitting the basil. Add 1 teaspoon dried tarragon leaves and 2 teaspoons fresh or freeze-dried chives. Whirl in blender until well blended.

#### Mogul Dressing

Combine in a bowl or jar 1/2 cup salad oil, 2 tablespoons tarragon wine vinegar, 1 teaspoon *each* dried basil leaves (crumbled) and Chinese Five Spice, 1/2 teaspoon *each* salt and lemon juice, 1 small clove garlic (minced or mashed), and 1/8 teaspoon *each* dry mustard and pepper. Beat with a fork or shake to blend. Let stand 2 hours or longer, then beat again and drizzle over peeled and thinly sliced tomatoes. You might arrange slices of avocado and cucumber with the tomatoes on the plate, and season with the same dressing. Makes 4 to 6 servings.

#### Sour Cream Dressing

Combine in a bowl 1/2 cup sour cream, 1/4 cup mayonnaise, 2 green onions (with about 2 inches of green tops) thinly sliced, 1 tablespoon minced parsley, and 2 teaspoons basil leaves, crumbled. Refrigerate for at least 30 minutes to blend flavors. Serve in a bowl at table to dress sliced tomatoes, which have been sprinkled lightly with salt and pepper. Makes enough for 4 servings.

## Pimiento-Tomato Gazpacho

2 slices firm white bread
1 large can (1 lb. 12 oz.) tomatoes
1 small can (2 oz.) pimientos, drained
1 or 2 cloves garlic, peeled
3 tablespoons olive oil
½ teaspoon salt
¼ teaspoon pepper
1 large cucumber, peeled

Break bread into pieces and put into the blender container; whirl until it makes fine crumbs; set aside or put into 4 to 6 individual soup bowls.

Into the blender container put half of the tomatoes; add the pimientos, garlic, olive oil, salt, and pepper. Blend at high speed until smooth. Then add remaining tomatoes and blend well; turn into a bowl. Dice the cucumber and add to the tomato mixture. Cover and refrigerate until well chilled.

At serving time, stir bread crumbs into tomato mixture. Or ladle soup over bread crumbs in serving bowls. Makes about 4 to 6 servings.

## Fresh Tomato Aspic

3 medium-sized tomatoes, peeled
1 package (3 oz.) lemon-flavored gelatin
1 cup boiling water
¼ cup lemon juice
¼ cup *each* chopped green pepper, green onion, and
   celery
⅛ teaspoon liquid hot pepper seasoning
   Crisp salad greens
   Avocado Dressing (recipe follows)

Cut each tomato in half and hold over a small bowl while squeezing gently to remove most of the seed pockets. Dice tomatoes and set aside.

Dissolve gelatin in boiling water; stir in lemon juice. Then stir in any juices from the tomato pulp after straining to remove seeds. Chill gelatin until syrupy. Stir in diced tomato, green pepper, onion, celery, and liquid pepper seasoning. Turn into a 4 or 5-cup salad mold and chill until firm.

To unmold, dip mold in hot tap water until edges of gelatin jiggle slightly (about 5 seconds). Cover mold with a serving plate; invert. Garnish with crisp greens. Pass dressing. Makes 6 servings.

**Avocado Dressing.** Peel and remove seed from 1 medium-sized avocado; put into a bowl and mash with a fork. Stir in 2 tablespoons lemon juice, 3 tablespoons milk, 2 tablespoons chopped chives (fresh or dried), 1/4 teaspoon *each* garlic salt and liquid hot pepper seasoning, and 1/2 teaspoon prepared mustard.

## Tomato and Sweet Onion Platter

3 large tomatoes
1 large sweet onion, thinly sliced
1 large cucumber, peeled and thinly sliced
1 large green pepper, seeded and cut into rings
½ teaspoon salt
¼ teaspoon pepper
½ teaspoon mixed salad herbs or basil leaves, crumbled
¼ cup salad oil
2 tablespoons white wine vinegar

Peel and cut tomatoes into 1/4-inch-thick slices; arrange in a single layer on a large rimmed platter. Separate onion slices into rings and arrange over tomatoes; top with the cucumber slices and then distribute pepper rings over all. Sprinkle with salt, pepper, and salad herbs. Drizzle salad oil and vinegar evenly over all. Cover and refrigerate for 2 to 4 hours. To serve, spoon some of the dressing over each portion. Serves 6.

## Basic Fresh Tomato Sauce

Use this sauce with spaghetti or to spoon on cooked meat, shellfish or vegetable.

    2 medium-sized onions, finely chopped
    4 cloves garlic, minced or mashed
    ⅓ cup olive oil or salad oil
    5 pounds (about 12 medium-sized) firm ripe tomatoes
    ½ cup minced green onion, including part of tops
    1 green pepper, seeded and chopped
 1½ teaspoons salt
    ¾ teaspoon pepper
    ¼ teaspoon anise seed, crushed
    1 tablespoon oregano leaves
    ¾ teaspoon rosemary leaves
    1 teaspoon paprika
      About 1¾ cups dry red wine

In a Dutch oven or large frying pan, cook onions and garlic in olive oil over medium heat until golden (about 15 minutes); stir occasionally.

Peel tomatoes as directed under *cutting and cooking*, page 88. Cut tomatoes into eighths and add to cooked onions along with the green onion, green pepper, salt, pepper, anise seed, oregano, rosemary, paprika, and red wine. Bring mixture to boil, stirring with a heavy wooden spoon to break up tomatoes. Cover, reduce heat, and simmer for 1 hour. Remove cover and boil until reduced to 8 cups.

You can use sauce while hot, let cool and refrigerate for 3 days, or freeze up to 4 months. If you freeze the sauce, divide it into 1, 2, or 4 cup-size portions to use according to the recipe you choose.

To reheat sauce (if frozen let thaw first), bring mixture to simmering over low heat, stirring occasionally. If sauce appears dry, blend in 2 to 4 tablespoons water or dry red wine. Makes about 2 quarts.

## Fresh Tomato Cocktail or Salad

    6 medium-sized tomatoes, peeled
    1 large clove garlic, minced or mashed
    ½ teaspoon basil leaves, crumbled
    ¼ teaspoon *each* oregano leaves, rosemary leaves, and
      thyme leaves, crumbled
    ¼ teaspoon salt
    ⅛ teaspoon pepper
    2 tablespoons *each* red wine vinegar and olive oil or salad
      oil
      Parsley sprigs or lettuce for garnish

For cocktails dice the tomatoes, discard most of the seeds; or for salad thinly slice tomatoes. Put diced or thinly sliced tomatoes in a bowl. Thoroughly blend the garlic, basil, oregano, rosemary, thyme, salt, pepper, vinegar, and oil. Pour over the tomatoes. Cover bowl and refrigerate for at least 4 hours or overnight.

At serving time, spoon out the diced tomato and juices into 6 or 8 seafood cocktail glasses (or small wine glasses) and garnish each serving with a sprig of parsley. For salads, lift tomato slices out of marinade and arrange on lettuce. Drizzle over some of the marinade. Makes 6 to 8 servings.

## Curry-Glazed Tomatoes with Rice

    2 tablespoons butter or margarine
    1 teaspoon curry powder
    1 medium-sized onion, chopped
    1 cup tomato juice
    ½ cup orange marmalade
    ½ teaspoon salt
    ¼ teaspoon pepper
    1 teaspoon ground cinnamon
    6 large tomatoes, peeled and cored
    3 cups hot cooked rice

Melt butter in a large saucepan over medium heat; add curry powder and onion; stir until onion is limp. Add tomato juice, marmalade, salt, pepper, and cinnamon; bring to a boil, stirring. Remove from heat and set aside.

Arrange whole tomatoes in a buttered baking dish (at least 2-qt. size) and pour sauce over all. (If you make this dish ahead, cover and refrigerate at this point.)

Bake, uncovered, in a 400° oven until tomatoes are just tender when pierced with a fork (about 20 minutes; 25 minutes if refrigerated). To serve, arrange tomatoes on a bed of rice in a serving dish and spoon sauce over all. Makes 6 servings.

## Bacon and Tomato Omelet

    1 large tomato, peeled
    6 eggs, lightly beaten
    ½ teaspoon *each* salt and basil leaves, crumbled
    ¼ teaspoon pepper
    2 tablespoons finely chopped parsley
    4 to 8 slices bacon
    1 medium-sized onion, chopped
    1 cup (about 4 oz.) shredded Cheddar cheese

Cut the tomato in half, and squeeze gently to remove some of the seeds. Then cut into slices and set aside.

In a bowl mix the eggs, salt, basil, pepper, and parsley; set aside.

In a 10-inch frying pan, cook bacon until crisp. Remove bacon from pan and drain; discard all except about 2 tablespoons drippings.

Place frying pan with 2 tablespoons drippings over medium-high heat; put in the onion, and cook, stirring often, until soft (3 to 5 minutes). Place tomato slices in a single layer over the onion; cook about 1 minute, then sprinkle cheese over tomatoes. Pour the egg mixture into the pan and use a large spoon to pull mixture from center to the sides of the pan, allowing eggs to flow into center. Cook just until bottom layer of egg is set (about 2 minutes). Set pan about 8 inches below heat in a broiler; heat until eggs are set on top (2 to 3 minutes). Place bacon on top of omelet and serve at once. Makes 4 servings.

## Spinach Baked Tomatoes

6 to 8 thick slices of large tomatoes or tomato halves,
   peeled
   Garlic salt and seasoned pepper
½ cup finely chopped onion
3 tablespoons melted butter or margarine
1 package (10 oz.) frozen chopped spinach
⅓ cup crushed bread stuffing mix
1 egg, slightly beaten
   Salt and pepper to taste
2 tablespoons grated Parmesan cheese

Arrange the tomato slices on a greased, shallow baking pan. Sprinkle lightly all over with the garlic salt and seasoned pepper. In a medium-sized frying pan, sauté the onion in butter for about 5 minutes. Add the spinach; heat and break it apart as it cooks until spinach is tender and the water is almost evaporated.

Remove from heat and stir in the stuffing mix and egg. Season with salt and pepper to taste. Mound part of the spinach mixture on each tomato slice, using 2 spoons or an ice cream scoop. Cover and refrigerate, if desired. Just before serving, sprinkle with cheese and bake, uncovered, in a 350° oven for about 15 minutes or until heated through. Makes 4 to 8 servings.

## Sautéed Green Tomato Slices

5 large or 8 small green tomatoes
½ cup all-purpose flour
1 tablespoon salt
⅛ teaspoon pepper
¼ teaspoon basil leaves, finely crumbled
1 tablespoon sugar
¼ cup (⅛ lb.) butter or margarine
1 cup sour cream
1 teaspoon sugar

Slice the tomatoes 1/2 inch thick and coat them on both sides with a mixture of the flour, salt, pepper, basil, and the 1 tablespoon sugar. Melt the butter in a large frying pan; sauté tomato slices slowly until browned on both sides. As the slices become browned and tender, remove to a serving platter and keep warm. When all the tomatoes are fried, reduce heat to low and add sour cream and the 1 teaspoon sugar to the pan. Heat, stirring, until a smooth sauce forms; pour over the sautéed tomatoes. Makes 6 servings.

## Hot Herbed Tomatoes

¼ cup (⅛ lb.) butter or margarine
1 teaspoon firmly packed brown sugar
½ teaspoon salt
   Dash pepper
½ teaspoon oregano leaves or basil leaves, crumbled
6 to 10 small, firm, ripe tomatoes (or 1 basket cherry
   tomatoes, peeled)
¼ cup finely chopped celery
2 tablespoons finely chopped parsley
2 tablespoons chopped chives or green onion

In a small frying pan with a cover, melt the butter. Stir in brown sugar, salt, pepper, and oregano. Set small tomatoes, if used, stem ends down. Cover and simmer gently for 5 minutes.

Add celery, parsley, and chives or green onion; turn small tomatoes over (or add cherry tomatoes, if used, at this point). Cover and cook about 5 minutes longer. Serve with sauce to spoon over tomatoes. Serves 6.

## Tomato Cups

6 medium-sized, firm, ripe tomatoes, peeled
½ green pepper, seeded and finely chopped
1 small onion, finely chopped
2 tablespoons butter or margarine
¾ cup dry bread cubes or unseasoned croutons
1 can (about 2¼ oz.) sliced ripe olives, drained
1 tablespoon sugar
¾ teaspoon salt
⅛ teaspoon pepper
2 tablespoons finely chopped parsley
1 egg

Cutting straight down with a small sharp knife, cut a hole around the center core of tomato 1 1/2 to 2 inches wide and to within 1/2 inch of the bottom. With a teaspoon, carefully remove pulp, leaving a shell at least 1/2 inch thick; discard the seeds and stem end, but chop and reserve firm center pulp. Turn tomatoes upside-down to drain.

In a frying pan, sauté the green pepper and onion in butter until limp (about 5 minutes). Add tomato pulp and bread cubes; cook, stirring for about 1 minute or until well blended. Remove from heat and stir in the olives, sugar, salt, pepper, parsley, and egg; mix well.

Pile filling mixture inside the tomato shells and arrange in a greased baking dish. (You can do this much several hours ahead.) Bake, uncovered, in a 350° oven for 30 to 40 minutes until heated through. Makes 6 servings.

## Pilaf-Stuffed Tomatoes

2 tablespoons olive oil
1 medium-sized onion, chopped
2 small cloves garlic, minced or mashed
½ cup long grain rice (uncooked)
1 cup regular strength chicken broth
¾ teaspoon salt
⅛ teaspoon pepper
½ teaspoon thyme leaves
6 medium-sized firm tomatoes, peeled and cored
2 tablespoons chopped parsley
⅓ cup grated Parmesan cheese

In a frying pan with a lid, heat 1 tablespoon of the oil over medium heat; add onion and garlic and sauté until limp and golden. Add rice and brown lightly. Pour in chicken broth and season with 1/2 teaspoon of the salt, pepper, and thyme. Cover, reduce heat, and simmer

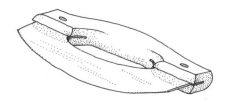

for 15 to 20 minutes until liquid is absorbed and rice is tender.

Cut a 1/2-inch-thick slice off core end of tomatoes; set aside. With a teaspoon, scoop out seeds and pulp; discard seeds and chop pulp along with the end slices. Sprinkle insides of tomato shells with remaining 1/4 teaspoon salt; turn upside down to drain.

Add chopped tomato and parsley to cooked rice. Place tomato shells in a shallow ovenproof baking dish; fill with rice mixture. Bake in a 375° oven for 10 minutes. Remove from oven; sprinkle filled tomatoes with cheese and the remaining 1 tablespoon oil. Return to oven and bake 10 minutes longer or until cheese melts and browns lightly. Makes 6 servings.

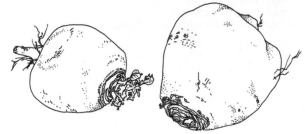

# TURNIPS

## Selection and Storage

**Selection:** Choose turnips that are small to medium-sized and firm with smooth skin. The tops should be fresh and green. Avoid turnips with too many leaf scars around the top with obvious fibrous roots.

**Buying:** Allow 1/2 pound for each serving.

**Storage:** Remove tops and store in a plastic bag in the refrigerator. Will keep 1 to 2 weeks.

## Cutting and Cooking

Remove thin layer of skin with a vegetable peeler. Wash; leave whole, dice, or cut in quarters, slices, or shred. (See page 56 for cooking turnip tops.)

**To cook turnips,** place in boiling salted water. Cook, covered, until just tender (20 to 30 minutes for whole; 5 to 10 minutes for cut pieces); drain.

## Seasoning and Serving

Turnips are milder in flavor than their relative, the rutabagas. They can be seasoned simply with salt, pepper, and melted butter; but they also go well with a variety of other seasonings ranging from sweet to hearty—and they are surprisingly good raw. Here are more ideas.

### Glazed Turnips

In a wide shallow pan, combine 2 cups thinly sliced turnips, 2 tablespoons *each* butter and water, and 1 tablespoon sugar. Cover and cook quickly, stirring frequently, until tender (about 5 minutes). Remove cover and sauté until glazed. Serves 4.

### Creamed Turnips

In a wide shallow pan, combine 2 cups thinly sliced turnips, 2 tablespoons *each* butter (or margarine) and water, and 1 tablespoon sugar. Cover and cook quickly, stirring frequently, until tender (about 5 minutes). Remove cover and add 1/4 cup whipping cream; simmer rapidly until slightly reduced. Makes 4 servings.

### Turnips with Bacon

Cook 3 cups diced turnips, covered, in 1/3 cup boiling salted water until tender (about 7 minutes); drain. Add 6 slices crumbled crisply cooked bacon, 1/2 cup whipping cream, and 1/3 cup shredded mild Cheddar cheese. Simmer rapidly, stirring, until sauce coats turnips. Makes 6 servings.

### Puréed Turnips

Peel and dice about 2 pounds turnips and cover with water to which 1 teaspoon each salt and sugar has been added. Cook until tender; drain and force through food mill or whirl in blender. Return to heat. Cook gently until the excess moisture has evaporated; add 1 cup medium white sauce made from your own recipe. Season with salt, pepper, ground nutmeg, and 1 tablespoon butter.

---

### Butter-Steamed Turnips

In a heavy wide frying pan or electric frying pan over high heat, melt 2 tablespoons butter or margarine, adding 4 cups thinly sliced turnips and 5 tablespoons water. Cover and cook, stirring occasionally, for 4 to 5 minutes or until tender. Season with salt and pepper. Makes 5 to 6 servings.

### Butter-Steamed Turnips in Soy Sauce

Prepare turnips as directed above for Butter-Steamed Turnips, omitting salt and pepper and adding 1 1/2 teaspoons soy sauce, or to taste. Makes 5 to 6 servings.

# Turnip Frittata

4 tablespoons (⅛ lb.) butter or margarine
1 medium-sized onion, finely sliced
1½ pounds turnips, peeled and coarsely shredded
1 clove garlic, crushed
3 eggs, slightly beaten
⅓ cup half and half (light cream)
½ cup shredded Cheddar cheese
   Salt and pepper to taste
⅔ cup croutons
   Paprika

Melt 3 tablespoons of the butter in a saucepan. Add onion and sauté until limp but not brown. Add turnip and cook for 4 minutes, stirring frequently. Cover and cook 4 minutes longer. Add garlic and remove from heat. Combine eggs, cream, and cheese; pour over the cooked turnips. Mix well. Add salt and pepper to taste.

Heat remaining 1 tablespoon butter in a heavy 9 or 10-inch frying pan until it begins to bubble. Add the turnip mixture and cook over medium heat until it begins to set around the edges. Sprinkle with the croutons, then sprinkle lightly with paprika. Bake, uncovered, in a 350° oven for 20 minutes or just until firm. Makes about 6 servings.

# Syrup-Glazed Turnips

2 pounds medium-sized turnips
   Boiling water
1¼ teaspoons salt
¼ cup regular strength chicken broth
1 teaspoon sugar
⅛ teaspoon *each* pepper and ground mace or nutmeg
2 tablespoons butter or margarine
3 tablespoons light corn syrup

Peel the turnips and cut in quarters. In a 9 or 10-inch frying pan (one that has a tight-fitting cover), bring about 1 inch water and the salt to a boil. Drop in turnip pieces and simmer for 2 to 3 minutes; drain. Add to turnips in pan the chicken broth, sugar, pepper, and mace. Cover and cook over medium heat for 6 to 8 minutes or until tender. Remove the cover during the last few minutes of cooking, if necessary, to evaporate the liquid. Add the butter and corn syrup and continue cooking over medium heat, uncovered, for 3 to 4 minutes. Makes 5 to 6 servings.

# Turnip Salad

2 bunches turnips (about 4½ lbs.), peeled
6 hard-cooked eggs
¼ cup chopped onion
2 cups sliced celery
1 cup mayonnaise
1 teaspoon prepared mustard
   Salt and pepper

Cook whole turnips following directions under *cutting and cooking*, page 92. Refrigerate until thoroughly cold, then cut in 1-inch cubes and put into a bowl. (You should have about 5 cups.) Chop 5 eggs and add to turnips in a bowl, reserving 1 egg for garnish. Add the onion, celery, mayonnaise blended with mustard, and salt and pepper to taste. Mix well and garnish with slices of egg. Makes about 8 servings.

# Alsatian Soup

¼ cup butter or margarine
1 medium-sized onion, chopped
1 shallot, chopped (optional)
1 cup sliced celery
½ pound mushrooms, sliced
1 medium-sized potato, peeled and diced
2 medium-sized turnips, peeled and diced
2 cans (about 14 oz. each) regular strength chicken broth
⅛ teaspoon marjoram leaves, crushed
2 teaspoons lemon juice
   Salt and pepper
⅓ cup half-and-half (light cream) or milk
   Chopped chives or parsley

Melt butter in a 3-quart saucepan over medium heat; add onion, shallot, celery, and mushrooms and cook until onion is soft. Remove and set aside about 1/2 the mushroom slices. Add potatoes, turnips, broth, and marjoram. Bring mixture to a boil, reduce heat, and boil gently, covered, for about 25 minutes or until potatoes and turnips are tender. Pour about 1/3 of the mixture at a time into a blender container and whirl until smooth; repeat until all mixture is puréed. (Or press vegetables through a wire strainer.) Return soup to the saucepan, stir in the lemon juice, reserved mushrooms, salt and pepper to taste, and cream. Reheat (do not boil). Garnish with chives before serving. Makes 4 to 6 servings.

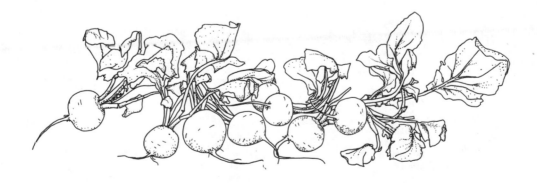

# INDEX